Inside SPICE

Other Reference Books of Interest by McGraw-Hill

Handbooks

ANTOGNETTI AND MASSOBRIO · *Semiconductor Device Modeling with SPICE*

BAKER · *C Mathematical Function Handbook*

BENSON · *Audio Engineering Handbook*

BENSON · *Television Engineering Handbook*

CHEN · *Computer Engineering Handbook*

COOMBS · *Printed Circuits Handbook*

DI GIACOMO · *Digital Bus Handbook*

DI GIACOMO · *VLSI Handbook*

FINK AND CHRISTIANSEN · *Electronics Engineers' Handbook*

HARPER · *Electronic Packaging and Interconnection Handbook*

HICKS · *Standard Handbook of Engineering Calculations*

INGLIS · *Electronic Communications Handbook*

JURAN AND GRYNA · *Juran's Quality Control Handbook*

KAUFMAN AND SEIDMAN · *Handbook of Electronics Calculations*

RORABAUGH · *Digital Filter Designer's Handbook*

TUMA · *Engineering Mathematics Handbook*

WAYNANT · *Electro-Optics Handbook*

WILLIAMS AND TAYLOR · *Electronic Filter Design Handbook*

Other

ANTOGNETTI · *Power Integrated Circuits*

BEST · *Phase-Locked Loops*

BUCHANAN · *CMOS / TTL Digital Systems Design*

BUCHANAN · *BiCMOS / CMOS Systems Design*

BYERS · *Printed Circuit Board Design with Microcomputers*

ELLIOTT · *Integrated Circuits Fabrication Technology*

HECHT · *The Laser Guidebook*

MUN · *GaAs Integrated Circuits*

SILICONIX · *Designing with Field-Effect Transistors*

SZE · *VLSI Technology*

TSUI · *LSI / VLSI Testability Design*

WATERS · *Active Filter Design*

WOBSCHALL · *Circuit Design for Electronic Instrumentation*

WYATT · *Electro-Optical System Design*

Inside SPICE

**Overcoming the Obstacles
of Circuit Simulation**

Ron M. Kielkowski

McGraw-Hill, Inc.

New York   San Francisco   Washington, D.C.   Auckland   Bogotá
Caracas   Lisbon   London   Madrid   Mexico City   Milan
Montreal   New Delhi   San Juan   Singapore
Sydney   Tokyo   Toronto

Library of Congress Cataloging-in-Publication Data

Kielkowski, Ron M.
  Inside SPICE : overcoming the obstacles of circuit simulation /
Ron M. Kielkowski.
    p.   cm.
  Includes bibliographical references and index.
  ISBN 0-07-911525-X (set)
  1. SPICE (Computer file)   2. Electric circuits—Computer
simulation.   I. Title.
  TK454.K48   1994   93-34238
  621.319'2'011353—dc20   CIP

International Edition 0-07-113424-7. Exclusive rights by McGraw-Hill Book Co.–
Singapore for manufacture and export. This book cannot be re-exported from the coun-
try to which it is consigned by McGraw-Hill. Not for resale in Australia, Canada,
Europe, Japan, the United Kingdom, and the United States of America. Export sales
may be made only by, or with the expressed consent of, the Publisher.

1 2 3 4 5 6 7 8 9 0   DOH/DOH   9 9 8 7 6 5 4 3

P/N 034627-5
PART OF
ISBN 0-07-911525-X

*The sponsoring editor for this book was Stephen C. Chapman, the editing supervisor
was Christine Furry, and the production supervisor was Suzanne W. Babeuf.
This book was set in Century Schoolbook by North Market Street Graphics.*

Printed and bound by R. R. Donnelley & Sons Company.

Hspice is a trademark of Meta-Software, Inc.
IS_Spice is a trademark of Intusoft, Inc.
Micro-Cap IV is a trademark of Spectrum Software.
Pspice is a registered trademark of MicroSim Corporation.
Rspice and Rgraph are trademarks of RCG Research, Inc.
MS-DOS is a registered trademark of Microsoft Corporation.

This book is printed on recycled, acid-free paper containing a minimum of 50% recycled de-
inked fiber.

To Walter Thomas Kielkowski

Contents

Preface

Although SPICE has been with us since 1971, use of the program during the past 10 years has exploded because of the personal computer. IBM's introduction of the desktop personal computer and the widespread availability of SPICE, both in the academic and commercial form, have made analog circuit simulation available to almost every circuit designer.

As a designer, the author struggled for many years with SPICE because, although the program holds an unlimited potential for aiding in circuit design, practical use of the program is limited by nonconvergence failures, timestep control errors, and numeric integration failures. These types of problems either stop the simulator or introduce error in the simulation output. Because of these failures, simulation results are either difficult to obtain or contain numerous inaccuracies.

Although these types of errors occur almost regularly, very little has been written about how to correct these simulation ills. The motivation for this text came from the lack of information available to designers on how to overcome these *obstacles of simulation*. Obtaining accurate high-quality simulation results requires learning how to overcome these simulation failures.

Someday, mathematicians and programmers will be able to produce a circuit simulator which always converges on the correct solution and always produces accurate results. Until then, as SPICE users, we must learn to overcome these limitations.

Book Organization

This book is written as both a tutorial and as a reference.

As a tutorial, the text includes a bound copy of the RSPICE™ circuit simulator and RGRAPH™ graphical postprocessor. Chapters 3, 4, 5, and 6 contain example circuits which should be simulated with RSPICE/RGRAPH. These circuits are integrated into the text and demonstrate the common failure mechanisms of SPICE as well as the actions required to correct the failure. The disk which accompanies the book contains all of the circuit files, organized by chapter, mentioned in the text.

As a reference, the book includes systematic, step-by-step procedures to correct common simulator failures. The book also includes multiple examples of simulation failures. These examples can be used to identify similar failures from the reader's simulations.

RSPICE/RGRAPH Disk

The accompanying RSPICE/RGRAPH disk contains the RSPICE circuit simulator and the RGRAPH graphical postprocessor. The disk also contains the circuit examples used throughout this text. An installation program automates the task of creating the proper directory structure, copying the files, and assigning the proper directory pointers.

Requirements

To install the programs and files, you will need approximately 2 MB of free space on your hard disk. The install program creates the following directories and subdirectories.

```
\RCGV33
\RCGV33\RSPICE
\RCGV33\CH2
\RCGV33\CH3
\RCGV33\CH4
\RCGV33\CH5
\RCGV33\CH6
\RCGV33\DEMO
```

The RSPICE and RGRAPH files are then copied to the appropriate directories.

Installation

We'll assume you're installing the software from a floppy disk. To install the programs, simply type:

```
X:install X: Y:
```

Where X is the floppy disk drive designator and Y is the hard disk drive designator. The installation program will complete the remaining installation chores.

Summary of Chapter Content

Chapter 1 is a brief overview of circuit simulation. Included in this chapter is a historical review of the development of CANCER, SPICE1, SPICE2, and SPICE3. The chapter continues with how computer simulation can be an aid in the design process and what is needed to simulate circuits productively and effectively.

Chapter 2 looks at how SPICE works, and shows how SPICE constructs the system equations from the elements of the input file, how SPICE iterates to a solution, how SPICE computes the DC bias, DC sweep, AC frequency sweep, and transient analysis solutions. Understanding the fundamental algorithms of SPICE is crucial to understanding how to correct the common simulator failure mechanisms.

Chapter 3 examines the problem of nonconvergence. Nonconvergence is one of the most common and most frustrating problems facing simulation users. But most nonconvergence problems can be overcome simply by using the options and controls found within SPICE. Chapter 3 looks at the causes of nonconvergence and presents systematic, step-by-step procedures which will eliminate (most) nonconvergence simulation failures.

Chapter 4 is the first of two chapters which focus on improving the accuracy of the transient analysis. In Chapter 4, the three numeric integration routines of SPICE are examined. Numeric integration failures introduce error into the simulation result, and SPICE does not have a routine which warns users of integration failure. As simulation users, we must learn to detect integration failure from the simulation result and take the appropriate corrective action.

Chapter 5 finishes the discussion on transient analysis by looking at the timestep control algorithms in SPICE. Like numeric integration, SPICE cannot detect timestep control failures. Users must learn to look for timestep failures in the simulation output. Chapter 5 examines different timestep control failures and illustrates the proper corrective action.

Chapter 6 examines the 34 options within SPICE. Each of the options are presented with a detailed explanation and suggested settings.

Appendix A compares the default option settings of several popular vendor-offered circuit simulators.

Throughout this text the program SPICE is referenced many times. For this book, SPICE refers specifically to Berkeley SPICE2G.6.

Ron M. Kielkowski

Acknowledgments

I would like to thank the many people who contributed to the development of this book. Let me begin by thanking Ed Oxner, whose decision not to write a book on SPICE served as the catalyst for this project; Dan Gonneau and Steve Chapman at McGraw-Hill for taking a chance on this first-time author; Brian Chastain and Art Mernone for reading the manuscript for clarity and correctness; Kim Hailey for his insights on Hspice; Charles Hymowitz for his help with IS_Spice; Wolfram Blume for his conversations on Pspice; Andrew Thompson for his assistance with Micro-Cap IV; and, of course, Ronald Rohrer, Donald Pederson, Larry Nagel, and the other Berkeley students and professors who contributed to the development of SPICE and whose efforts changed the face of analog design.

Finally, I would like to thank my beautiful wife Christine and my family, without whose love and infinite patience this book would never have become reality.

Ron M. Kielkowski

What Is SPICE?

SPICE, and specifically SPICE2G.6, is an engineering design tool which falls into the category of general purpose analog circuit simulators. Just some of the tools which fall into this same category include SPICE3, PSPICE, HSPICE, IS_SPICE, MICRO-CAP IV, RSPICE, SABER, and many others.

Computer simulation can be a powerful supplement to traditional design techniques. In most design strategies, simulation can be an aid in the initial design development, during the breadboarding phase, and during debugging and diagnostic phases. For some circuits, initial design theories must be tested before circuit design begins. With simulation, circuit blocks may be represented as behavioral elements and simulated in a functional form. Behavioral elements allow designers to test circuit theory without the time involved in developing transistor and component-level descriptions of each circuit function.

For many circuits, breadboarding is impossible because of excessive circuit complexity, layout-specific parasitic effects, or, as in the case of integrated circuits, both effects. For these types of circuits, simulation may be the only avenue to investigate the circuit behavior before building a working prototype.

For most circuits, component-value variation will have a direct effect on circuit performance and product yield. With simulation, designers can effectively predict the performance of a circuit as one or more circuit variables are changed.

For all of these reasons, computer simulation and SPICE are playing an increasingly important role in electronic circuit design.

What Is a SPICE-Compatible Simulator?

Almost all circuit simulators read a *circuit file* which describes the types of elements in the circuit and the type of analysis to be performed. After processing the circuit file, the simulator performs the desired analysis and generates the output in either tabular or graphical form.

Of all the analog circuit simulation tools available, the overwhelming majority of them are SPICE-like or SPICE-compatible. SPICE-like means a simulator is capable of producing an analysis result similar to the SPICE result for a given circuit, although the SPICE-like simulator may not be able to read a *standard* SPICE circuit file. SPICE-compatible means a simulator will read a SPICE circuit file, perform the desired analysis and produce the output result in standard SPICE2G.6 form. Table 1.1 illustrates the SPICE compatibility of several vendor-offered simulators.

The Birth of SPICE

Ironically, the man most responsible for the development of SPICE wasn't interested in circuit design or circuit simulation. In 1968, Ron Rohrer, a junior faculty member at the University of California, Berkeley, and an authority on circuit *optimization,* was assigned to teach a class in network synthesis. But Rohrer was convinced that synthesis was of little practical use. Rohrer was much more interested in circuit optimization and optimization techniques. Circuit optimization involves performing multiple simulations on a circuit to study the change in output with a change in one or more circuit variables. Up to this time, Rohrer's work with optimization had been stifled by the lack of a fast, efficient circuit simulator. Because of the unavailability of simulation tools which were appropriate for use in circuit optimization, Rohrer decided to replace the course on circuit synthesis with a class on circuit simulation. Ultimately, Rohrer was hoping to develop a new circuit simulator for his work in optimization.[1]

TABLE 1.1   Vendor-Offered Simulator Compatibility

SPICE-compatible	SPICE-like
Hspice	Saber
Pspice	Micro-Cap II, III
IS_Spice	
Rspice	
Micro-Cap IV	
Accusim	
Analog Workbench	

CANCER

In the class on circuit simulation, Rohrer and a dozen students assembled a nonlinear circuit simulator which was to become the foundation for SPICE. This first simulator was christened CANCER[2] (Computer Analysis of Nonlinear Circuits Excluding Radiation) by a student named Larry Nagel.

CANCER was capable of DC operating point, DC sweep, transient sweep, and AC frequency sweep analyses, the same analysis types available today in SPICE. CANCER simulations might include resistors, capacitors, inductors, and two types of nonlinear devices, junction diodes and bipolar transistors. Diodes were modeled with the Shockley diode equations, and bipolar transistors were modeled with the Ebbers-Moll transistor equations. While CANCER was an outstanding tool for its day, the program's "lifetime" was limited because the solution routines could handle no more than 400 components and/or 100 circuit nodes.

CANCER wasn't the only simulator available in 1970.[3] Other programs of the time included IBM's ECAP and ECAP-II. BIAS[4] and SLIC[5] (Simulator for Linear Integrated Circuits) originated from early work on CANCER. Autonetics, a division of Rockwell, developed TRAC[6] (Transient Radiation Analysis by Computer Program) in 1968. The TRAC program was the basis for Motorola's TIME and MTIME[7] programs. TRAC was also the foundation for Berkeley's SINC program.

SPICE1

During the early 1970s, Nagel continued improving the CANCER program. In 1971, this improved version, named SPICE1 (Simulation Program with *Integrated Circuit Emphasis*), was released into the public domain. Because Berkeley distributed the program with almost no charge, SPICE1 quickly became an industry standard simulation tool. SPICE1 offered several improvements over CANCER. The bipolar transistor model was changed to the Gummel-Poon[8] model. JFET and MOSFET devices were added with the Shichman-Hodges[9] model. SPICE1 also offered a new approach to modeling known as macromodeling. With macromodels, engineers could describe portions of a circuit in the form of relocatable circuit templates (subcircuits).

During this time, the rapid development of the integrated circuit industry fueled the work on SPICE. In many ways, integrated circuits were very different from board-level circuits. Many IC problems could not be examined with traditional design techniques. Because of this, computer simulation of integrated circuits proved to be an invaluable design tool.

Because of the push from the integrated circuits industry, however, many of the algorithms in SPICE were optimized for these types of circuits. This is a limitation which has lived on to this day. While SPICE can be used to simulate almost any type of electronic circuit, board-level circuits and circuits using inductors or transformers require extra care because many of the SPICE algorithms may not be well suited to these types of circuits.[10]

SPICE2

The next major release of the program came in 1975 with the introduction of SPICE2. SPICE2 offered significant improvements over SPICE1 or CANCER, including a new equation formulation for voltage-defined elements (inductors and voltage-controlled voltage and current sources). Both the accuracy and speed of transient analysis were improved with the development of two dynamic timestep control algorithms and a multi-order implicit integration scheme. SPICE2 device model improvements were added to keep pace with the changing device technologies. From 1975 through 1983, Berkeley continued improving and upgrading the SPICE2 program. In 1983, SPICE2 version G.6 (SPICE2G.6) was released to the public domain. SPICE2G.6 was to be the last fortran version of SPICE released by the university. (SPICE2G.6 is still available from Berkeley.)

SPICE3

CANCER and both generations of SPICE were written in fortran source code. With the increasing use of Unix-based workstations, Berkeley made the decision to rewrite the SPICE2 program in C. The new C version of the program was known as SPICE3.

SPICE3 was to be a superset of SPICE2, including all of the analysis types and device models of SPICE2 as well as new features such as improved device models, voltage- and current-controlled switches, pole-zero analysis, and a graphical postprocessor for viewing simulation results. Unfortunately, much of the enthusiasm (and research money) available in the early '70s had disappeared, and the overwhelming task of converting some 22,000 lines of fortran code to C was left to a handful of students. Because of the enormity of the task, the first release of SPICE3A.1 contained dozens of bugs and coding errors. Worse yet, somewhere in the translation between fortran and C, much of the SPICE2G.6 functionality was lost. SPICE3 became a program which was not backward-compatible with SPICE2G.6. As of this writing, SPICE3F.2 is the latest release of SPICE3. Although Berkeley continues to improve the program, SPICE3 is not yet completely backward-compatible with SPICE2.

Some of the features which have not been written into SPICE3 include the ability to run multiple circuit netlists from a single file, polynomial-controlled dependent sources, polynomial capacitors and inductors, on-line resistor temperature coefficients, circuit topology error checking, and temperature sweep analyses.[11] While *work-arounds* are available for most of these features, SPICE2 users may need to rewrite old circuit netlists to make them compatible with SPICE3; this includes rewriting old macromodels to make them SPICE3-compatible. Because of the hundreds of SPICE2 macromodels available today, and because of the large installed base of SPICE2-compatible simulators, the lack of backward compatibility is probably the overriding reason SPICE3 has not replaced SPICE2 as the industry-standard circuit simulator.

Although it may not have all of the features of SPICE2, SPICE3 offers several technical advantages to SPICE2. SPICE3 is written in modular C code which is easier to modify than the fortran of SPICE2. SPICE3 demonstrates superior convergence characteristics to the SPICE2 algorithms. In rewriting the SPICE3 device models, several SPICE2 errors were discovered and corrected. As Berkeley continues adding enhancements and SPICE2 compatibility to the program, the day will come when SPICE3 will replace SPICE2. But since the vast majority of us are using a SPICE2-based simulator, this text will focus on SPICE2.

From 1980 up through today, several vendor-offered versions of SPICE appeared. Some of the better-known simulators which got their start in this time frame include Meta-Software's HSPICE, IntuSoft's IS_SPICE, Spectrum Software's MICRO-CAP, and MicroSim's PSPICE. All of these were developed from the original SPICE2 framework. Figure 1.1 illustrates the functional relationship between SPICE2G.6 and the vendor-offered simulators. While many other SPICE-based programs exist, these four represent the best-known simulators. Although a handful of vendors offer products based on SPICE3, the overwhelming majority of SPICE-like simulators are still based on SPICE2G.6.

Why Simulate?

Why should you simulate your circuits? What are the benefits of simulation? Too many times you hear the answer, "Circuit simulation replaces breadboarding." Nothing could be further from the truth! *Simulation does not replace the breadboard; simulation complements the breadboard.* Many things can be learned from the breadboard that cannot be learned from simulation, and simulation may reveal many things not readily learned in the lab.

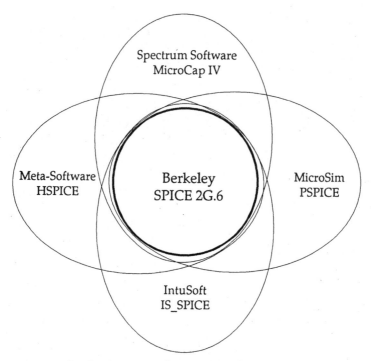

Figure 1.1 Graphical representation of SPICE-compatible simulators. (*Reprinted from* Successfully Simulating Circuits With SPICE. *Used with permission.*)

Although thousands of reasons can be given for simulating circuit designs, the majority of them fall into one of the following four categories.

Verifying design theories

Simulation offers the ability to quickly test circuit design theories before a single wire is soldered or mask fabricated. Verifying the theory of a design can be done at different levels, behavioral models, macro-models, and circuit component (transistor) models.

Behavioral models express circuit blocks in mathematical relationships. In behavioral form, simulating an entire system response becomes realizable. Behavioral blocks are easy to construct, and simulate hundreds or thousands of times faster than a circuit at the component level. Behavioral blocks represent the highest level of the simulation hierarchy.

Macromodels express circuit blocks in the form of simplified equivalent circuits. Macromodel circuit elements may include real circuit components and ideal circuit components (dependent voltage and current sources). Macromodels simulate slower than behavioral models but much faster than transistor-level models. Because of their simplified representation, macromodels do not give the fine detail or accuracy of a transistor-level simulation.

A transistor-level simulation is simply a simulation of the components in a circuit. This type of simulation offers the most accuracy of any analysis but often with a stiff run-time penalty. A transistor-level simulation represents the lowest level of the simulation hierarchy.

In verifying circuit theories, designers often begin simulation with behavioral models. As the design progresses and more accuracy is required, macromodels replace the behavioral models. If even more accuracy is required, transistor-level models may replace the macromodels. In many cases, a simulation may be made up of behavioral models, macromodels, and transistor-level models, all in a single run. The ability to simulate at these different levels offers users a fast, efficient means to test the theory of a design.

Circuit performance and yield analysis tool

Simulation allows the designer to quickly test a circuit over a variety of conditions including temperature variations, element value variations, and power supply variations. Once a circuit netlist has been established, circuit parameters (including temperature, element values, or power supply levels) may be altered and resimulated. The ability to alter circuit parameters offers designers a fast, efficient means of testing the circuit operation under a variety of operating conditions.

In the real world, as opposed to the simulation world, resistors, capacitors, inductors, transistors, and other electrical component values vary from part to part. Even under the tightest process control, most common electrical components have part-to-part variations of 1 percent to 5 percent or more. Designers recognize this and try to develop designs which are insensitive to these variations. If the yield of a circuit depends on the component value variations, simulation may be useful in estimating circuit yields.

Circuit simulation offers users a fast, efficient means to test a circuit's performance characteristics as one or more of the component values change. By performing multiple simulations and observing the output of each simulation, component variation limits can be established to guarantee a functional circuit. Figures 1.2a and b illustrate how multiple simulations may be used to characterize the performance of a circuit.

Many of the vendor-offered SPICE packages now offer Monte Carlo and worst-case analyses. Monte Carlo and worst-case analyses auto-

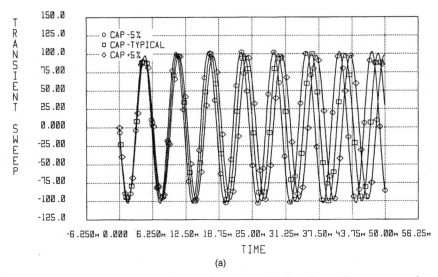

Figure 1.2(*a*)   Simulation is an efficient means to study the effects of component-value variations.

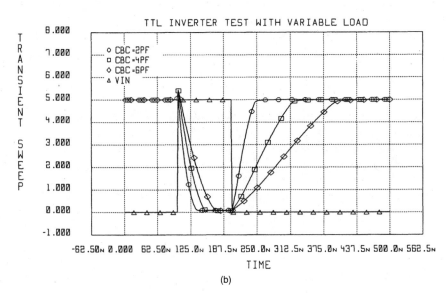

Figure 1.2(*b*)   Simulation is an efficient means to study the effects of component-value variations.

matically vary circuit components as multiple simulations are performed. Monte Carlo and worst-case analyses enhance a user's ability to measure circuit performance over a number of changing circuit parameters.

Evaluating circuit vulnerability

Computer simulation gives the designer a chance to test a circuit for vulnerability prior to being assembled. Sneak-path conditions, signal race conditions, and power dissipation/power overload conditions can all be quickly and easily investigated through the use of simulation. While many of these conditions could be examined on a breadboard, when circuit simulation models are forced beyond their performance limits, they don't smoke, smolder, smell, fuse, or explode like their breadboard equivalents. Simulation allows a designer to examine the circuit without the risk of damage to the circuit or the designer!

Failure analysis/diagnosis/prognosis

As a failure-analysis tool, circuit simulation has limits. The most obvious limit is the lack of ability to *predict* layout-dependent parasitic behaviors. SPICE simulates components as if they were ideal, noninteracting elements. For example, a common problem with CMOS (complementary metal-oxide-silicon) circuits is the phenomenon known as latch-up. Latch-up occurs when an N-channel and P-channel transistor pair are fabricated in close proximity. Because of the p- and n-type silicon layers in the transistor pair, a pair of parasitic bipolar transistors is also formed. Latch-up occurs when the parasitic bipolar transistors begin to conduct current. If an N-channel and P-channel transistor pair are entered in SPICE, the latch-up effect will not be observed *unless* the parasitic bipolar transistors are included in the netlist. Figure 1.3 illustrates the parasitic bipolar transistors formed in P-well CMOS circuits. Trying to use SPICE to predict this type of failure is futile.

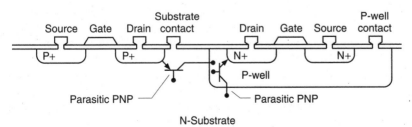

Figure 1.3 Parasitic bipolar transistors formed in a P-well CMOS inverter.

While SPICE will not predict layout-dependent parasitic effects before they are known, once these effects are known and the parasitic elements entered in the netlist, SPICE can be used to help determine a solution to the problem. In the example of the CMOS latch-up, once the parasitic bipolar transistor has been identified as the cause of the failure, a designer may discover that adjusting the beta (Hfe) of the transistors may reduce or eliminate the latch-up effect. Once a parasitic element has been identified as the cause of a problem, simulation can be a fast, efficient means to determine how to correct the problem.

These are just some of the reasons that so many designers are turning to simulation. Simulation does not replace the breadboard. Often, the type of information you obtain from the simulation is not the same as the information gathered from a real prototype, and the converse is also true. Used together, the breadboard and the circuit simulator form an indispensable design aid.

Which Simulator Is Best?

This is a question without an answer. Asking which simulator is best is like asking which op-amp is best, which bipolar transistor is best, or which solder is best. You can order a copy of IntuSoft's IS_SPICE for $99, or you can order a Cray version of Meta-Software's HSPICE for $120,000. Obviously there are differences between the two programs but there are also several similarities. Figure 1.1 illustrates the functional relationship between SPICE2G.6 and several of the vendor-offered simulators. Functionally, most of the SPICE-compatible simulators are supersets of SPICE, which means that a circuit netlist which runs in SPICE2G.6 will also run in HSPICE, IS_SPICE, MICRO-CAP IV, or PSPICE, but the converse is not true. Circuits which use HSPICE-specific functions and features will not run in SPICE2G.6, nor will they run in PSPICE, MICRO-CAP, or IS_SPICE.

The question you should ask is, "Which simulator is best suited *for my design needs?*" Of course, SPICE2G.6 can still be ordered from Berkeley for $150 (source code only). But many of the vendor-offered programs offer advantages and capabilities over and above Berkeley SPICE. Some of the features added in the vendor-offered tools include improved device models, Monte Carlo and worst-case analysis, pole-zero analysis, network analysis, scaleable parameters and functions, automatic optimization, device model libraries, and hotline support. While many of the added features of one vendor-offered tool may overlap another, many features are unique to any SPICE-compatible simulator. Look for the features which will best meet your design needs. Only you can determine which simulator is best for the way you design.

A word about price

As with most free-market commodities, the price you pay for a simulator depends on the features and capabilities added to the program, the number of copies a vendor expects to sell (personal computer software usually sells for less than workstation software because the PC market is much larger), and the speed of the machine on which you run the software. The last reason may seem a little odd. Many software companies have decided that running a simulator on a faster machine means obtaining the results sooner, or being more productive with their software. Because of this, many software vendors adjust the price of the tools based on the speed of the computer. Shop around. You may find you can save a bundle depending on the type of computer you're using.

What Do You Need to Simulate?

In the most basic form, simulation requires a text editor to create the input files and a simulation program. A minimum machine recommendation is a 386 PC (with the coprocessor). As far as a simulation program and editor, the MS-DOS editor (EDIT) is an easy way to create your first input files.

If you are using a PC for simulation, the simulator you start with should be a 32-bit version of SPICE. SPICE uses 64-bit math throughout the program. Many of the early PC versions of SPICE were 16-bit programs. To move a single-node voltage or branch current required four memory transfers. A 32-bit program moves data almost twice as fast as a 16-bit version of SPICE; because of this, a 32-bit version of SPICE executes much quicker than a 16-bit version. While the 16-bit versions of SPICE will run on any PC (from the pcXT through 486s), the 32-bit programs will only run on 386sx, 386dx, 486, and higher CPUs. The RSPICE and RGRAPH programs which accompany this text are 32-bit programs. Ask your simulation vendor if their program is a 32-bit or 16-bit version.

But the preceding paragraph describes only the basics. If you start with just a simulator and editor, you will quickly realize that accurate simulation requires accurate component models and accurate device model parameters. One of the most important features any of the vendor-offered SPICE versions can add is a library of common electrical components. Models of bipolar transistors, diodes, op-amps, and thousands of other devices can be stored in a library and called into the simulator as needed. Of course, you can develop your own device models, but developing accurate component models and model parameter sets can be a costly, time-consuming process.

Whether you obtain device models from a library or develop them yourself, before you can obtain accurate, reliable, meaningful results, you need to know how to *use your simulator*. Using your simulator means learning how to assemble input files in the syntax of SPICE, learning how to avoid the problems of nonconvergence, timestep control, and numeric integration inaccuracies. A common trap new SPICE users fall into is assuming the simulation output is always correct. Often SPICE will fail to converge on a solution or, worse yet, will converge on the wrong solution! Numeric integration and timestep control problems can lead to inaccurate results. Even though the SPICE program is over 20 years old, we are still learning which numeric algorithms work well and which ones don't.[11]

The final ingredient for high-quality simulations is user education. Learning how to overcome nonconvergence problems, learning to decide whether the simulation output is accurate or erroneous, and learning to diagnose simulation ills are the lessons which must be mastered to achieve accurate, efficient simulations.

These three elements—a good simulator, accurate models, and good user skills—are like the legs of a tripod; no one element is more important than any other. Without a good simulator, accurate device models, and good user skills, simulation results are simply imaginary numbers.

The obstacles of simulation

While the preceding paragraphs proclaim how wonderful SPICE is, in truth, the average user will have many problems when using the program. SPICE is not a foolproof design aid. For many years, the author had the job of helping the engineers of a large design community learn to use SPICE. Although several SPICE-based packages were in use, including SPICE2G.6, many simulations still suffered from nonconvergence problems, timestep control anomalies, and numeric integration errors. *For many designers, the obstacles of simulation are the inability to both determine the cause of and correct an unexpected simulation result.*

This book is appropriately titled, *Inside SPICE: Overcoming the Obstacles of Circuit Simulation*. The obstacles of simulation are things which make the simulator fail, like nonconvergence; things which make the simulator produce the wrong answer, like timestep control problems; and things which make the simulator produce unexpected results, like numeric integration instabilities. Many designers don't realize SPICE can make mistakes, much less know how to correct a mistake if it occurs. These are the real obstacles to circuit simulation, and these are the problems this text attempts to address, identify, and correct.

Summary

SPICE can be an indispensable design/analysis tool, but, like any tool, SPICE has limitations. Even the finest tools are ineffective in the hands of an inexperienced user. SPICE is a complicated design tool. You cannot simply toss a netlist and device models into a simulator and expect correct results. Like a craftsperson skillfully making a fine piece of furniture, a simulation user will carefully select the simulator options which are appropriate for a given circuit and analysis type. Accurate simulation results don't happen by accident. Both the device models and the simulator options you select will determine the speed, the accuracy, and the efficiency of your result.

This text focuses specifically on the algorithms of SPICE2G.6. Since the majority of SPICE-like simulators are based on the SPICE2G.6 algorithms, almost everything presented in this text will apply directly to the SPICE-like simulators. While many of the vendors have added modifications to the original 2G.6 algorithms, none of the SPICE-like simulators has completely eliminated nonconvergence problems, timestep control problems, or numeric integration problems. For these reasons, it is important for all SPICE users to understand why these problems arise and how they can minimized or eliminated. All of the solutions presented in this text are in the form of standard user-selectable SPICE controls, and most of the SPICE-like simulators have these same controls. Teaching SPICE users how to tailor the simulator's control parameters to enhance the speed and accuracy of the simulation result is the goal of this text; the lessons learned here can be applied to every SPICE-like simulator available today.

References

1. R. H. Rohrer, "Circuit Simulation—The Early Years," *IEEE Circuits and Devices Magazine,* May 1992.
2. L. W. Nagel, "Computer Analysis of Nonlinear Circuits, Excluding Radiation (CANCER)," *IEEE Journal of Solid-State Circuits,* August 1971.
3. A. Vladimirescu, "SPICE—The Third Decade," IEEE Bipolar Circuit and Technology Meeting, September 1990.
4. W. J. McCalla, "BIAS-3: A Program for the Nonlinear DC Analysis of Bipolar Circuit Transistors," *IEEE J. Solid-State Circuits,* Vol. SC-6, February 1971.
5. T. E. Idleman, "SLIC—A Simulator for Linear Integrated Circuits," *IEEE J. Solid-State Circuits,* Vol. SC-6, August 1971.

6. E. D. Johnson, "Transient Radiation Analysis by Computer Program (TRAC)," Autonetics Div., North American Rockwell Corp., Tech. Rep., Harry Diamond Labs., June 1968.
7. L. W. Nagel, "SPICE2: A Computer Program to Simulate Semiconductor Circuits," Electronics Research Laboratory Rep. No. ERL-M520, University of California, Berkeley, 1975.
8. H. K. Gummel and H. C. Poon, "An Integral Charge-Control Model of Bipolar Transistors," *Bell Systems Tech. Journal,* Vol. 49, May 1970.
9. H. Shichman and D. A. Hodges, "Modeling and Simulation of Insulated-Gate Field-Effect Transistor Switching Circuits," *IEEE J. Solid-State Circuits,* Vol. SC-3, 1968.
10. T. L. Quarles, "Analysis of Performance and Convergence Issues for Circuit Simulation," Memo. UCB/ERL M89/42.
11. *SPICE Version 3F User's Guide.*

Chapter

2

Understanding Circuit Simulation

The SPICE Engine

SPICE has been available for over 20 years, yet many simulation users understand very little about how SPICE works. Many engineers know much more about the parameters of their automobile engine than they do about the parameters of the SPICE engine. Test yourself. How many cylinders does your engine have? How much horsepower does the engine produce? How many valves per cylinder do the heads have? How many forward gears are in the transmissions? Many of you will answer these without a moment's hesitation, and very few of us would buy a car if we did not know these basic characteristics of the automobile.

But how would you answer the same type of questions regarding your circuit simulator? How many solution algorithms does SPICE use to solve the circuit equations? How many timestep control algorithms does SPICE offer for transient analysis? Which numeric integration routine does SPICE use? What is the smallest voltage or current SPICE can resolve? What is the upper limit on resistance in SPICE? How would you do on these questions?

In this chapter, these and other questions about the simulator will be answered. Here, the internal workings of SPICE will be exposed. You will learn how the simulator works, how it represents circuit elements in the system matrix, and how the matrix is filled and solved. This chapter is going to look under the hood of a simulation vehicle, at the engine of SPICE.

Under the Hood

Many people learn about automobiles by simply raising the hood, looking, pointing, and asking questions. Unfortunately, software doesn't lend itself to this same type of troubleshooting. So not surprisingly, how circuit simulation works is not widely understood.

Although it may seem so at times, there is nothing magical about SPICE. SPICE will not generate a single solution which could not be generated by paper and pencil calculations. The SPICE program simply automates the calculations.

The System Equations

SPICE starts an analysis by writing a set of nodal equations which describe the elements in the circuit. Figure 2.1 illustrates a simple resistive circuit. The nodal equations for this circuit may be constructed by summing the currents leaving each of the three circuit nodes, V_1, V_2, and V_3. Figure 2.2a and Eq. 2.1 describe the current leaving the V_1 circuit node.

$$-3 \text{ amps} + \frac{V_1 - V_2}{5 \text{ ohms}} = 0 \qquad (2.1)$$

Figure 2.2b and Eq. 2.2 describe the current leaving the V_2 circuit node.

$$\frac{V_2 - V_1}{5 \text{ ohms}} + \frac{V_2}{10 \text{ ohms}} + \frac{V_2 - V_3}{5 \text{ ohms}} = 0 \qquad (2.2)$$

Figure 2.2c and Eq. 2.3 describe the current leaving the V_2 circuit node.

$$\frac{V_3 - V_2}{5 \text{ ohms}} + \frac{V_3}{10 \text{ ohms}} = 0 \qquad (2.3)$$

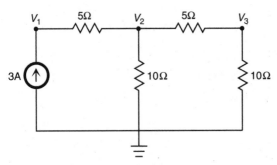

Figure 2.1   A simple linear resistive circuit.

With simple algebraic manipulations, Eq. 2.1–2.3 may be written in the form of Eqs. 2.4–2.6, respectively.

$$.2*V_1 - .2*V_2 \qquad = 3 \qquad (2.4)$$

$$-.2*V_1 + .5*V_2 - .2*V_3 = 0 \qquad (2.5)$$

$$-.2*V_2 + .3*V_3 = 0 \qquad (2.6)$$

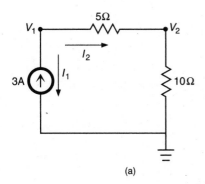

(a)

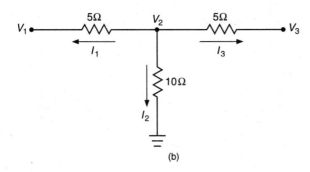

(b)

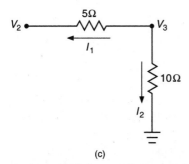

(c)

Figure 2.2(a), (b), and (c) Summing the branch currents to determine the nodal equations.

Equations 2.4–2.6 are the nodal system equations which describe the circuit of Fig. 2.1. These three equations have three unknown voltages V_1, V_2, and V_3. Gaussian elimination is one of the best-known methods for solving simultaneous equations and can be used if we rewrite Eqs. 2.4–2.6 in matrix form.

$$
\begin{bmatrix}
.2 & -.2 & 0 \\
-.2 & .5 & -.2 \\
0 & -.2 & .3
\end{bmatrix}
*
\begin{bmatrix}
V_1 \\
V_2 \\
V_3
\end{bmatrix}
=
\begin{bmatrix}
3 \\
0 \\
0
\end{bmatrix}
\tag{2.7}
$$

Equation 2.7 is the matrix form of the system equations. Applying forward-elimination to Eq. 2.7 yields Eq. 2.8. (Readers are encouraged to derive Eq. 2.8 from 2.7 just for the fun of it.)

$$
\begin{bmatrix}
.2 & -.2 & 0 \\
0 & .3 & -.2 \\
0 & 0 & .25
\end{bmatrix}
*
\begin{bmatrix}
V_1 \\
V_2 \\
V_3
\end{bmatrix}
=
\begin{bmatrix}
3 \\
3 \\
3
\end{bmatrix}
\tag{2.8}
$$

The value of voltages V_1, V_2, and V_3 may be determined by applying back-substitution to Eq. 2.8 resulting in:

$$
V_3 = 3/.25 \qquad = 12 \text{ volts} \tag{2.9}
$$

$$
V_2 = \frac{[.2*(V_3) + 3]}{.3} = 18 \text{ volts} \tag{2.10}
$$

$$
V_1 = \frac{[.2*(V_2) + 3]}{.2} = 33 \text{ volts} \tag{2.11}
$$

(Author's note: It took the author approximately 12 minutes to work through this example; it took SPICE exactly .09 seconds to calculate the same result!)

Elements in the Matrix

For many circuits, SPICE uses simple nodal analysis techniques to determine the circuit voltages. But the solution algorithm becomes much more complicated when nonlinear devices and charge-storage elements are included in the circuit. Nonlinear and charge-storage elements must be reduced to simplified equivalent circuits before being entered in the system equations. These simplifications are required because the system matrices only accept linear I-V relationships. To understand how SPICE develops the matrix, the linear I-V characteristic of each circuit element must be found.

The matrix

The system equations in SPICE are represented in a set of matrices. The equations characterize the linear representation between the voltage and current for every element in the circuit. The system matrices are shown in Fig. 2.3.

The voltage array is the solution array and represents the node voltages of the circuit. When performing a simulation, SPICE tries to determine the node voltage values which satisfy Kirchhoff's voltage and current laws for the circuit. The solution voltages found in the simulation output come directly from the voltage array.

The current array is one of the known values of the circuit and represents the independent branch currents generated from current sources and active devices. During simulation, SPICE determines the branch currents from the current source settings or from the previous voltages applied to the active device terminals.

The conductance array is another known value of the circuit and represents the *linear* relationship between voltage and current for every element in the circuit. The values in the conductance array are determined from the elements of the circuit. Nonlinear elements such as diodes and transistors, and charge-storage elements such as capacitors and inductors, are represented in the conductance array by their linear equivalent circuits. The entries in the current and conductance arrays are used to determine the next set of solution voltages.

Resistors

Circuit resistors only appear in the conductance array. The value in the conductance array represents the inverse of the resistance value. Figure 2.4 represents the I-V characteristic of a resistor and the resistor's equivalent conductance value. The resistor's value changes neither with applied voltage nor time, so the value of conductance which represents the resistor remains constant for the duration of the simulation.

$$
\begin{bmatrix}
G_{11} & G_{12} & G_{13} & \cdots \\
G_{21} & G_{22} & \cdots & \\
G_{31} & \vdots & & \\
\vdots & & &
\end{bmatrix}
*
\begin{bmatrix}
V_1 \\
V_2 \\
V_3 \\
\vdots
\end{bmatrix}
=
\begin{bmatrix}
I_1 \\
I_2 \\
I_3 \\
\vdots
\end{bmatrix}
$$

Figure 2.3   The system matrices used in SPICE.

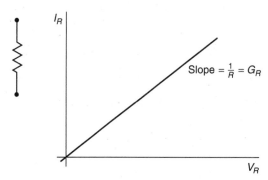

Figure 2.4   The I-V characteristics of a resistor.

Nonlinear elements

Representing diodes, transistors, and other nonlinear devices in the system equations is slightly more complicated than simple resistors. The reason for the complication is that nonlinear devices are defined by current-voltage relationships which change in response to the DC bias of the device. This is a problem for the system equations because the matrices represent *linear* relationships ($G * V = I$). The conductance array represents the linear I-V relationship for every element in the circuit.

The way around this problem is to break the nonlinear I-V relationships into many smaller linear approximations. Figure 2.5a illustrates a simple nonlinear function. Figure 2.5b illustrates the same function expressed as several linear pieces. Any nonlinear function may be expressed as a series of linear (piecewise linear) approximations. As the number of linear segments increases, the accuracy of the approximation increases. SPICE uses this same technique to simulate nonlinear circuit elements.

During the simulation, SPICE transforms the nonlinear elements into simple linear equivalent circuits. The linear equivalent circuit is determined from the voltage bias of the device. At each step in an analysis, SPICE uses the linear equivalent circuit to represent the device *as it is operating at that voltage bias.*

Figure 2.6a illustrates the I-V characteristics of a diode. If the solution (voltage) is close to the voltage V_b, the only part of the diode I-V characteristic of interest to the simulator is the I-V characteristics close to V_b. In this case, the nonlinear characteristics of the diode may be represented by a straight-line approximation equal to the tangent of the diode characteristic.

Figure 2.6b illustrates a straight-line model of the diode. The straight-line model is determined from the tangent of the diode I-V characteris-

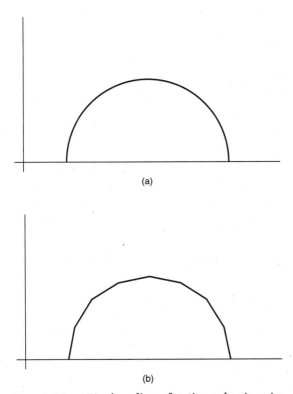

(a)

(b)

Figure 2.5(a) and (b)   A nonlinear function and a piecewise linear approximation of the function.

tics at the voltage bias V_b. The straight-line model intersects the I_d axis at Ieq and has a slope equal to G_d, the dynamic diode conductance. This straight-line model is also known as the *linearized* diode model. The technique of building the linear model is known as *linearization*. While the original nonlinear diode characteristics are expressed by Eq. 2.12, Eq. 2.13 represents the linearized diode equation.

$$I_d = I_s\left[\exp\left(\frac{qV_d}{NKT}\right) - 1\right]$$ (2.12)

$$I_d = G_d{}^*V_d + I_{eq}$$ (2.13)

Figure 2.7 illustrates an equivalent circuit representation of the linearized diode equation (2.13). The circuit in Fig. 2.7 is the equivalent circuit SPICE uses to represent diodes in the system equations. This equivalent circuit is known as the linear diode model. In SPICE, diodes are modeled by a conductance (resistance) in parallel with a current

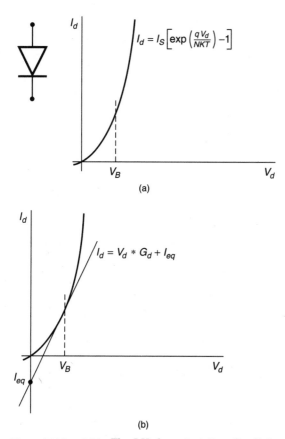

$$I_d = I_S \left[\exp \left(\frac{qV_d}{NKT} \right) - 1 \right]$$

(a)

$$I_d = V_d * G_d + I_{eq}$$

(b)

Figure 2.6(a) and (b) The I-V characteristics of a diode and a linear approximation to the diode curve at the voltage V_b.

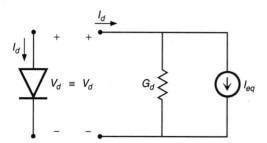

Figure 2.7 A diode and the linear equivalent circuit of the diode.

source. During a simulation, the conductance value G_d ($1/G_d$ is also known as the small-signal resistance or the dynamic resistance) will be stored in the conductance array, and the equivalent current I_{eq} will be stored in the current array. In turn, these values will be used to compute a new set of solution voltages. At each solution point, the diode current and conductance values are computed and stored within the system equations.

When discussing the linear diode model, the distinction between the linear model and the *small-signal* diode model should be made. Although the linear diode model is similar to the small-signal diode model, the two are not the same. The linear model of the diode contains two elements, the dynamic conductance (G_d) of the diode and the equivalent current (I_{eq}). The small-signal diode model also contains two elements, the dynamic resistance ($1/g_d$) and the small-signal capacitance (c_d). The linear diode model may or may not contain the small-signal capacitance (c_d) (depending on whether a transient analysis is being performed), and the small-signal diode model does not contain the equivalent current (I_{eq}). Figure 2.8a illustrates the linear diode model, whereas Fig. 2.8b illustrates the small-signal diode model.

Transistors are represented by extending the linear diode model. For bipolar, JFET, and MOSFET transistors, the linear model is established by using the same linearization technique previously described for the diode. But in the case of transistors, the linearization process must account for every pair of device terminals. For the bipolar, the base-emitter, collector-emitter, and collector-base terminals are linearized. For the MOSFET, the gate-source, drain-source, source-drain, and source-bulk terminals are linearized. Once the linearized models are found, the conductance and equivalent current values are stored in the system equations.

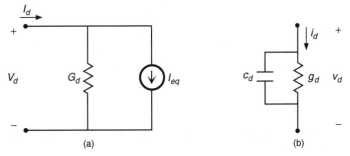

(a)   (b)

Figure 2.8(a) and (b)   The linear equivalent circuit of a diode and the small-signal model of a diode.

Charge-storage elements

While understanding the nonlinear devices is made much easier by drawing the I-V characteristics of the device, understanding the charge-storage elements is more difficult because the I-V characteristics cannot be drawn as a simple two-dimensional graph. For capacitors and inductors, the I-V characteristics change with the applied voltage, current, and time. But, like the linear and nonlinear devices, the capacitor and inductor values must be stored in the system equations.

In a small-signal frequency sweep, inductors and capacitors represent linear elements, but during a transient time sweep, capacitors and inductors possess a definite nonlinear I-V characteristic *which is a function of time.* To visualize the I-V characteristics of a capacitor would require a three-dimensional graph, with one axis representing voltage, one representing current, and one representing time.

Figure 2.9 illustrates a simple capacitor circuit. At time $T=0$, the switch closes and current flows onto the capacitor plates. Over a period of time, the capacitor voltage rises to the battery potential, and the current flowing into the capacitor decreases to zero. Figure 2.10a illustrates the relationship of the capacitor current and time. Figure 2.10b illustrates the relationship of the capacitor voltage and time.

To visualize the I-V characteristics of the capacitor, imagine Figs. 2.10a and b placed together on a common time axis. Figure 2.11a illustrates the superposition of Figs. 2.10a and b on a common axis. Notice that the common axis in Fig. 2.11a and b is marked in discrete time intervals. The linear I-V relationship of the capacitor can be found by drawing a line between the capacitor current on the Y axis and the capacitor voltage on the X axis. The resulting line represents the linear I-V relationship of the capacitor *at that moment in time.* If you follow this procedure for each timepoint, you will witness the I-V relationship change over time. This effect is illustrated in Fig. 2.11b. Figure 2.12 illustrates Fig. 2.11b as viewed along the time axis.

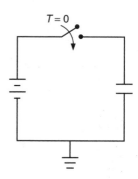

Figure 2.9 A simple time-domain circuit.

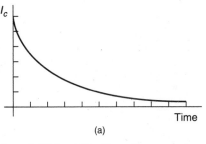

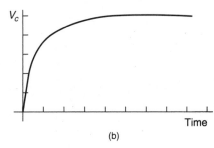

(a)   (b)

Figure 2.10(a) and (b)   Capacitor current vs. time and capacitor voltage vs. time for Fig. 2.9 circuit.

During transient analysis, the solution points are discrete moments in time. At each solution timepoint, the *numeric integration* routines determine the linear I-V relationship for capacitors and inductors. The linear relationship is expressed graphically in Fig. 2.13 and mathematically in Eq. 2.14.

$$I_c = G_c{}^*V_c + I_{eq} \qquad (2.14)$$

In SPICE, charge-storage elements are represented with simplified linear equivalent circuits. (For the capacitor, the circuit represents the current-voltage relationship expressed in Eq. 2.14.) These equivalent circuits are known as the *companion models*. Figures 2.14a and b illustrate the companion model of the capacitor and inductor. The conductance and current source value (or resistance and voltage source value for an inductor) of the companion model are recalculated at each new transient timepoint by the numeric integration algorithms. Once determined, the elements of the companion models are stored in the system matrices. At each new solution timepoint in a transient analysis, SPICE computes new companion models for each capacitor and inductor in the circuit.

SPICE simulates the behavior of electrical components by representing each element as a simple linear equivalent circuit. This is known as *linearizing* the circuit. As the simulation progresses, SPICE recalculates new linearized equivalent circuits for each nonlinear and charge-storage element in the circuit. As the circuit voltages and currents change, the linearized models change to reflect the nonlinear behavior of the circuit.

Matrix construction by inspection

In the section on the system equations in this chapter, you saw a simple circuit transformed into a set of nodal equations. The nodal equations were then transformed into a set of system equations and put in

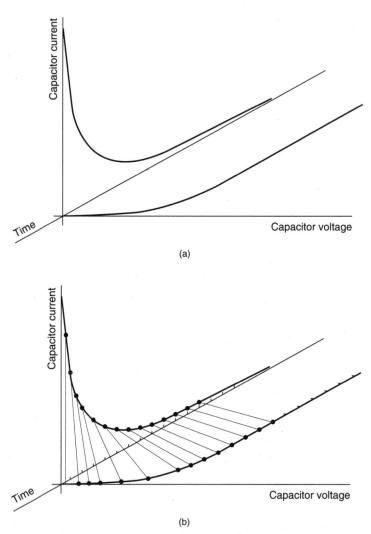

(a)

(b)

Figure 2.11(a) and (b)   Capacitor current and voltage vs. time and the I-V characteristics of discrete timepoints.

matrix form. SPICE basically uses this same procedure to build the system equations and circuit matrix. To save time, though, SPICE uses a shortcut to develop the system matrix. This shortcut allows SPICE to build the matrix as each element is read from the input file. This shortcut is known as *matrix construction by inspection.* Matrix construction by inspection builds the system matrices and identifies the location of each element in the matrices as soon as the nodes connected to the element are defined.[1]

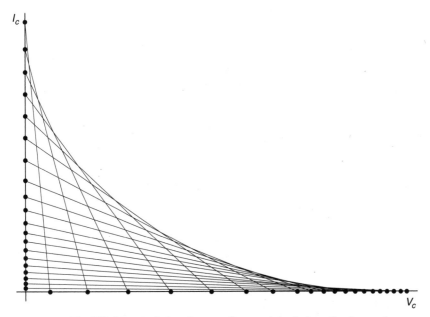

Figure 2.12   The I-V characteristics of a capacitor as viewed along the time axis.

Matrix construction by inspection builds the system matrix with the help of predefined element templates. The templates describe the position in the matrix for the conductance and current values of a particular device. The templates for the resistor, the current source, and the diode are shown in Fig. 2.15*a, b,* and *c.* Every element which can be simulated in SPICE has an *element template.* Even the four-terminal MOSFET transistors have a predefined element template. The template for the MOSFET is shown in Fig. 2.16. The system matrix in SPICE expands as more elements are read from the input file. When SPICE reads the element from the input file, the element template describes the positions in the conductance and current arrays for the

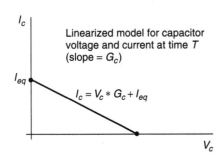

Linearized model for capacitor voltage and current at time T (slope $= G_c$)

$I_c = V_c * G_c + I_{eq}$

Figure 2.13   The linearized model for a capacitor voltage and current at time T.

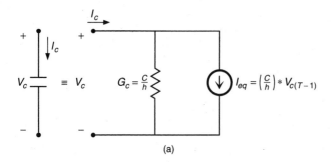

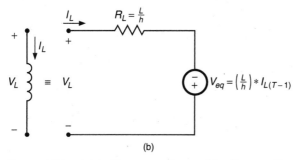

Figure 2.14(a) and (b) The companion models of a capacitor and an inductor.

device. The template also determines whether the arrays need to be expanded to add additional nodes for new elements. By the time SPICE reads the last element from the circuit file, all of the circuit components have been assigned proper locations in the system matrix.

Once the templates for an element are known, matrix construction by inspection may be applied to almost any circuit. As a simple example, examine Fig. 2.1. A SPICE netlist for the circuit shown in Fig. 2.1 is shown in Fig. 2.17.

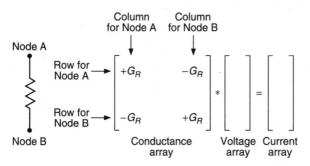

Figure 2.15(a) The template of a resistor.

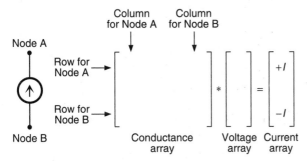

Figure 2.15(*b*)   The template of a constant current source.

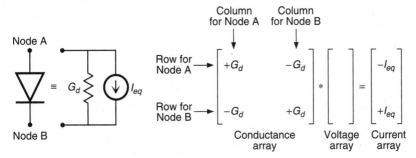

Figure 2.15(*c*)   The template of the diode equivalent circuit.

As SPICE reads the current source I_1, the current source template assigns the node connections to the rows and columns in the matrices which correspond to the nodes of the source (node 1 and node 0). As prescribed by the template, SPICE creates an initial 2×2 system matrix as shown in Fig. 2.18*a*. SPICE continues by reading the resistor R_1. Using the resistor template, the matrix is increased by one additional node as shown in Fig. 2.18*b,* and the conductance values corresponding to the resistor are assigned to the conductance array.

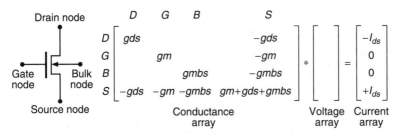

Figure 2.16   The template of a four-terminal MOSFET.

```
                    Example Circuit

I1 0 1 3A
R1 1 2 5_OHMS
R2 2 0 10_OHMS
R3 2 3 5_OHMS
R4 3 0 10_OHMS
.END
```

Figure 2.17   SPICE netlist for Fig. 2.1.

The conductance of a 5-ohm resistor is .2 mhos, and the value of the resistor is placed in the conductance array as prescribed by the template. The R_2 resistor is added to Fig. 2.18c. Notice how *overlapping* template entries are added to existing entries. The R_3 resistor is added to Fig. 2.18d, and R_4 is added to Fig. 2.18e. Figure 2.18e is a 4×4 matrix which corresponds to the four nodes in the circuit (nodes 1, 2, 3, and ground). But since *ground* is defined as the reference voltage and is equal to zero, the matrix row and column which correspond to ground may be eliminated. Figure 2.18f illustrates the final matrix after the elimination of the ground row and column. Notice that Fig. 2.18f is an exact match to Eqs. 2.4–2.6, and all this was achieved without writing a single nodal equation!

SPICE uses matrix construction by inspection to initialize the system equation matrix. Matrix construction by inspection is a fast and efficient manner to construct the system equations. SPICE uses matrix

$$
\begin{array}{c}
\quad\quad\text{Node 0}\quad\text{Node 1}\\
\begin{array}{c}\text{Node 0}\\[3em]\text{Node 1}\end{array}
\begin{bmatrix} & \\ & \\ & \end{bmatrix}
*
\begin{bmatrix} V_0 \\ \\ V_1 \end{bmatrix}
=
\begin{bmatrix} -3 \\ \\ 3 \end{bmatrix}
\end{array}
$$

(a)

$$
\begin{array}{c}
\quad\text{Node 0}\quad\text{Node 1}\quad\text{Node 2}\\
\begin{array}{c}\text{Node 0}\\\text{Node 1}\\\text{Node 2}\end{array}
\begin{bmatrix} & & \\ & .2 & -.2 \\ & -.2 & .2 \end{bmatrix}
*
\begin{bmatrix} V_0 \\ V_1 \\ V_2 \end{bmatrix}
=
\begin{bmatrix} -3 \\ 3 \\ 0 \end{bmatrix}
\end{array}
$$

(b)

Figure 2.18(a) and (b)   An example of matrix construction by inspection.

Node 0   Node 1   Node 2

$$
\begin{array}{c}
\text{Node 0} \\
\text{Node 1} \\
\text{Node 2}
\end{array}
\begin{bmatrix}
.1 & & -.1 \\
& .2 & -.2 \\
-.1 & -.2 & .3
\end{bmatrix}
*
\begin{bmatrix}
V_0 \\
V_1 \\
V_2
\end{bmatrix}
=
\begin{bmatrix}
-3 \\
3 \\
0
\end{bmatrix}
$$

(c)

Node 0   Node 1   Node 2   Node 3

$$
\begin{array}{c}
\text{Node 0} \\
\text{Node 1} \\
\text{Node 2} \\
\text{Node 3}
\end{array}
\begin{bmatrix}
.1 & & -.1 & \\
& .2 & -.2 & \\
-.1 & -.2 & .5 & -.2 \\
& & -.2 & .2
\end{bmatrix}
*
\begin{bmatrix}
V_0 \\
V_1 \\
V_2 \\
V_3
\end{bmatrix}
=
\begin{bmatrix}
-3 \\
3 \\
0 \\
0
\end{bmatrix}
$$

(d)

Figure 2.18(c) and (d)   An example of matrix construction by inspection.

construction by inspection to develop the system equations as the circuit components are read from the input file. By the time the last element has been read from the file, the system matrices have been defined completely.

The Matrix Solution

Once the conductance and current arrays have been filled, SPICE must solve the node voltage values in the voltage array. SPICE has two solution algorithms, one for linear circuits and one for nonlinear circuits. In this text a *linear circuit* is defined as one which contains only linear elements, voltage sources, and current sources. A *linear element* is defined as a passive element that has a linear voltage-current relationship.[2] For a DC and transient analysis, linear elements include resistors and linear-dependent voltage and current sources. For an AC small-signal analysis, linear elements include resistors, capacitors, inductors, and linear-dependent voltage and current sources.

Linear analyses

If a circuit contains only linear elements, SPICE uses the computer equivalent to Gaussian elimination to solve the matrix. The algorithm SPICE uses is known as LU decomposition. For the purpose of understanding, LU decomposition is an efficient computer method of per-

Node 0   Node 1   Node 2   Node 3

$$
\begin{array}{c}
\text{Node 0} \\
\text{Node 1} \\
\text{Node 2} \\
\text{Node 3}
\end{array}
\begin{bmatrix}
.2 & & -.1 & -.1 \\
& .2 & -.2 & \\
-.1 & -.2 & .5 & -.2 \\
-.1 & & -.2 & .3
\end{bmatrix}
*
\begin{bmatrix}
V_0 \\
V_1 \\
V_2 \\
V_3
\end{bmatrix}
=
\begin{bmatrix}
-3 \\
3 \\
0 \\
0
\end{bmatrix}
$$

(e)

Node 1   Node 2   Node 3

$$
\begin{array}{c}
\text{Node 1} \\
\text{Node 2} \\
\text{Node 3}
\end{array}
\begin{bmatrix}
.2 & -.2 & \\
-.2 & .5 & -.2 \\
& -.2 & .3
\end{bmatrix}
*
\begin{bmatrix}
V_1 \\
V_2 \\
V_3
\end{bmatrix}
=
\begin{bmatrix}
3 \\
0 \\
0
\end{bmatrix}
$$

(f)

Figure 2.18(e) and (f)   An example of matrix construction by inspection.

forming Gaussian elimination to solve a linear set of equations.[1] An example of Gaussian elimination was used earlier in this chapter in Eqs. 2.7–2.11.

Nonlinear analyses

SPICE uses Gaussian elimination to solve the circuit equations when only linear elements are present. But, with the introduction of one or more nonlinear elements, SPICE must use a different solver algorithm. The second solver is a nonlinear solution technique known as the Newton-Raphson algorithm.

Iterating to a solution.   If a circuit contains only linear elements, the circuit equations are algebraic in nature. Equation 2.15 is an example of a simple algebraic equation.

$$7x + x = 32 \tag{2.15}$$

Equation 2.15 may be solved with simple algebraic manipulations as shown in Eqs. 2.16 and 2.17.

$$8x = 32 \tag{2.16}$$

$$x = 4 \tag{2.17}$$

But with the introduction of one or more nonlinear elements in the circuit, the circuit equation becomes transcendental in nature. Equation 2.18 is an example of a transcendental equation.

$$ln(x) + x = 32 \qquad (2.18)$$

Equation 2.18 cannot be solved with simple algebraic manipulations. Instead, transcendental equations are often solved with an iterative guessing technique. Begin by rearranging Eq. 2.18 into Eq. 2.19.

$$f(x) = ln(x) + x - 32 \qquad (2.19)$$

Construct a table as shown in Fig. 2.19a. Equation 2.19 may be solved by making a series of iterative guesses until, at the proper value of X, the $f(x)$ result is equal to or very close to zero. The author's succession of guesses is shown in Fig. 2.19b.

When one or more nonlinear devices is introduced into the circuit, the system equations become transcendental equations and cannot be solved with simple Gaussian elimination. In this case, a nonlinear solution technique known as the Newton-Raphson algorithm is applied to the system matrix. The Newton-Raphson algorithm, a method of successive approximations, is an iterative approach to solving a set of nonlinear equations. This algorithm starts the iterative process with an initial guess and finds the solution through a series of successive guesses.

Equation 2.20 is the Newton-Raphson formula.

$$X_{n+1} = X_n - \frac{F(X_n)}{F'(X_n)} \qquad (2.20)$$

A simple example of the Newton-Raphson formula might be helpful here. To apply the Newton-Raphson algorithm solution algorithm to

x	f(x)

(a)

x	f(x)
1	-31
10	-19.7
20	-9
30	1.4
29	.36
28	-.66
28.75	.10
28.725	.08
28.675	.03
28.650	.005
28.645	.000021

(b)

Figure 2.19(a) and (b) Iterating to the solution of a transcendental equation.

Eq. 2.18, rewrite Eq. 2.18 in the form illustrated in Eq. 2.19. This will be the function $F(x)$. The solution to this equation is the value of X which makes the function $F(x)$ equal zero. To apply the Newton-Raphson algorithm, the derivative of Eq. 2.19 must be found. Equation 2.21 represents the derivative $F'(x)$.

$$f'(x) = \frac{1}{x} + 1 \qquad (2.21)$$

The resulting Newton-Raphson formula is shown in Eq. 2.22.

$$X_{n+1} = X_n - \frac{ln(Xn) + Xn - 32}{\left(\dfrac{1}{Xn} + 1 \right)} \qquad (2.22)$$

Starting the iterative process with an initial guess of $X_n = 1$ results in the table of values shown in Fig. 2.20. Notice that after the fourth iterative value, the values of X_n and X_{n+1} remain constant. The Newton-Raphson iterative process begins with an initial guess and terminates when the difference between successive guesses falls to zero. When the Newton-Raphson algorithm has found the exact solution, the value predicted by any additional iterations remains unchanged. In Fig. 2.20, the fourth and fifth iterations are identical and match the solution calculated earlier in Fig. 2.19b.

Terminating the iterations. SPICE uses the Newton-Raphson algorithm to solve the circuit equations when one or more nonlinear devices are entered in the circuit. SPICE starts with an initial guess for every node voltage in the circuit and begins iterating. With each successive iteration, a new set of node voltages is predicted. The solution routine monitors the node voltage of the present iteration and the previous iteration value. Ideally, at the exact solution, the node voltage between successive iterations should be identical, or the difference between iterative voltage values should be zero. But, because of the way digital computers represent numbers, saying when two numbers are exactly equal can be difficult (because of round-off errors). Because of this dif-

x	f(x)
1	16.5
16.5	28.47
28.47	28.645
28.645	28.645
28.645	28.645

Figure 2.20 The Newton-Raphson iterations on a transcendental equation.

ficulty, SPICE monitors the difference between iterative node voltage values and compares the difference with a predefined error tolerance. When the difference between iterative voltage values is less than the error tolerance for every node of the circuit, SPICE terminates the iterative process for that solution point.

Nonconvergence in SPICE. In addition to the error tolerance limits, SPICE limits the total number of iterations each analysis type is allowed to process. If the iterative node voltages have not satisfied the error tolerance requirements before SPICE exceeds the iteration limit, SPICE aborts the simulation and proclaims the infamous nonconvergence error message.

Each different analysis type has a different limit on the number of iterations allowed. Each analysis iteration limit may be reset by the user with the .OPTIONS statement. Chapter 3 will explain how to set each of the iteration limits.

Newton-Raphson example. The Newton-Raphson algorithm is an iterative process which allows SPICE to quickly determine the node voltage values of the circuit. To see how SPICE applies the Newton-Raphson algorithm to a circuit, look at the circuit shown in Fig. 2.21.

The diode of the circuit is characterized by the I-V equation shown in Eq. 2.23.

$$I_d = 1pA*[exp(40*V_d) - 1] \qquad (2.23)$$

From this, the nodal equation for this one-node circuit may be written by summing the currents leaving the V_1 node. The nodal equation is shown in Eq. 2.24.

$$F(V_d) = 0 = -5 + \frac{V_d}{2} + 1pA*[exp(40*V_d) - 1] \qquad (2.24)$$

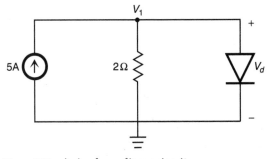

Figure 2.21   A simple nonlinear circuit.

Equation 2.24 forms our circuit function $F(V_d)$. To apply the Newton-Raphson formula to this circuit, the derivative of Eq. 2.24 must be found. The derivative of Eq. 2.24 is shown in Eq. 2.25.

$$\frac{dF(V_d)}{dV_d} = 0 + \frac{1}{2} + 40\text{pA}^*\exp(40^*V_d) \tag{2.25}$$

The iterative equation for this circuit is shown in Eqs. 2.26 and 2.27.

$$X_{n+1} = X_n - \frac{F(X_n)}{F'(X_n)} \tag{2.26}$$

$$V_{d+1} = V_d - \frac{-5 + .5^*V_d + 1\text{pA}^*[\exp(40^*V_d) - 1]}{.5 + 40\text{pA}^*\exp(40^*V_d)} \tag{2.27}$$

Starting with an initial guess of the voltage at 1 volt, 14 iterations are required to reach .7291 volts. The iteration voltage values are shown in Table 2.1. Notice the 14th iteration in Table 2.1; both V_d and V_{d+1} are .7291. The iterative procedure was set to terminate when the voltage-to-voltage iterations were equal to within four significant digits.

In this example, the initial voltage was chosen to be 1 volt. If a different starting voltage had been used, the iterative procedure will still iterate to .7291 volts, but depending on the initial voltage, the iterative process may require more or less iterations.

Table 2.2 illustrates the number of iterations required to iterate to .7291 volts for a number of different starting voltages. Notice that the number of iterations changes substantially with the initial guess. *The Newton-Raphson algorithm converges quickly to a solution if the initial guess is close to the exact solution.* The previous statement is important

TABLE 2.1   Iterative Voltage Values for Diode Circuit

V_d	V_{d+1}	Iteration #
1.00000	0.975001	1
0.975001	0.950002	2
0.950002	0.925005	3
0.925005	0.900015	4
0.900015	0.875041	5
0.875041	0.850113	6
0.850113	0.825309	7
0.825309	0.800838	8
0.800838	0.777250	9
0.777250	0.755885	10
0.755885	0.739445	11
0.739445	0.730983	12
0.730983	0.729186	13
0.729186	0.729119	14

TABLE 2.2   Iterations Required vs. Starting Voltage

Starting voltage	# Iterations
.6 V	119
.65 V	22
.7 V	5
1 V	14
2 V	54
3 V	94

because if SPICE does not achieve convergence within the allowed number of iterations, the simulation will terminate with a nonconvergence failure.

In the disk included with this text, the ch_2 subdirectory contains a file named DIODE.EXE. This simple program contains the diode circuit of Fig. 2.19 and the Newton-Raphson algorithm. The program prompts the user for a starting voltage, then prints the resulting iterative voltage values. This program was used to generate the entries in Table 2.2. Experiment with the DIODE.EXE program and different starting voltage values. Confirm the entries in Table 2.1.

Example summary. SPICE uses the Newton-Raphson algorithm as the nonlinear solution engine. The iterative process begins with an initial guess for every node voltage in the circuit. The iterative process continues until either the solution values between iterations are identical or nearly identical, or the number of iterations exceeds the allowed limit. If the latter occurs, SPICE prints a nonconvergence warning message and terminates the simulation.

Because there is a limit on the number of iterations allowed to find the solution, selecting an initial voltage close to the exact solution may be very important in overcoming nonconvergence problems. Setting an initial node voltage will be discussed in detail in Chap. 3.

SPICE Operation Overview

SPICE begins an analysis by reading elements from the input file. Using *matrix construction by inspection* and a set of *predefined element templates,* the system equations are described in a set of linear matrices. Once the system matrix has been defined, SPICE performs the analyses described in the input file commands.

DC operating point analysis

Every analysis begins with a DC operating point calculation. The DC operating point calculation establishes the DC bias of the circuit. In addition, the operating point voltages are used as the initial condition for all other sweep analyses.

During the DC operating point analysis, only the DC currents and voltages are calculated. The circuit capacitors are modeled as ideal open circuits, and the circuit inductors are modeled as ideal short circuits.

To calculate the DC operating point, the matrix entries representing voltage and current sources are set to their proper source values. With this complete, SPICE makes an initial guess of the node voltage for every other node in the circuit and stores the guess in the voltage array.

When the voltage array has been filled with initial values, SPICE calls a routine known as LOAD. The LOAD routine uses the initial node voltage guess to calculate the equivalent current and conductance (linearized circuit models) for each nonlinear element in the circuit. (The LOAD routine also calculates the companion model for the charge-storage elements during the transient analysis.) The LOAD routine stores the equivalent currents in the current array and the equivalent conductances in the conductance array as prescribed by the element templates.

When all the devices have been stored in the current and conductance arrays, SPICE passes the arrays to the Newton-Raphson solver. The Newton-Raphson solver uses the matrix form of Eq. 2.18 to calculate a new set of node voltages. This new set of node voltages is the first set of iterative voltage values, and these values replace the previous voltages in the voltage array. Once the new node voltages are stored, SPICE again calls the LOAD routine to calculate new linearized circuit models based on the new node voltage values. The resulting conductance and current values are stored in the current and conductance arrays, and the iterative solution process continues.

This cycle of performing a Newton-Raphson iteration, calling the LOAD routine to calculate new current conductance values based on the previous iterative voltage values, and then performing another Newton-Raphson iteration, occurs over and over again. The iterative cycle continues until either all the node voltages and branch currents match the previous iterative values, or until the number of Newton-Raphson iterations exceeds the allowed number of iterations.

To calculate the DC bias point, SPICE uses an initial guess for the node voltages in the circuit. The initial guess is used as a starting point in the Newton-Raphson iterations. Normally, the DC bias point solution requires somewhere between 10 and 500 iterations. If the solution is found, SPICE prints the bias voltages in the output file. If the solution is not found before the allowed number of iterations, SPICE prints a nonconvergence failure message in the output file and aborts the simulation.

DC sweep analysis

The DC sweep analysis is simply a series of DC operating point calculations. In the DC sweep analysis, a source voltage or current is stepped over a range of values. At each step in the progression, SPICE performs the same DC bias point calculation described previously. But in the sweep analysis, each step may require from 2 to 50 iterations to obtain the solution just for that point!

Because the sweep analysis may require hundreds or thousands of solution points, the simulation usually takes much longer to complete than the DC operating point calculation. But the authors of SPICE recognized something which would help speed the sweep analysis. Because the sweep consists of many evenly spaced steps, the node voltage changes between any two steps is usually small. Knowing this, the authors of SPICE decided to use the previous solution point node voltages as the initial guess to the Newton-Raphson iterations for the next solution point. This technique works well and significantly reduces the number of iterations required to solve each step in the DC sweep.

As the sweep progresses, at each new solution point SPICE saves the voltages and currents named on the .PRINT statement in memory. When the analysis is complete, the output routines in SPICE print the analysis results in either tabular or graphical form.

AC frequency sweep analysis

Like the DC sweep analysis, the AC frequency sweep starts by calculating the DC bias of the circuit. Once the bias has been established, the nonlinear large-signal transistor and diode models are replaced by their linear small-signal models. The small-signal models are determined from the DC bias point of the circuit.

The small-signal models replace the large-signal models because in SPICE the AC frequency sweep is defined as a small-signal linear analysis. This means that during the AC frequency sweep, distortion, clipping, saturation, and other nonlinear effects are ignored after the bias point has been established. To SPICE users who are unaware of the use of small-signal models during the AC frequency sweep, the output results are sometimes surprising. (For example, if you perform an AC frequency sweep on an op-amp with an open-loop gain of 1e6, a supply voltage of +15VDC to −15VDC, and an input voltage magnitude of 1VAC, SPICE will predict an output voltage magnitude of 1,000,000 volts. To see the op-amp clip at the power supply voltages, a DC sweep analysis or a transient times sweep analysis would be required.)

Small-signal analysis also implies the use of complex quantities like voltage phase and magnitude. For SPICE to simulate complex quantities, the voltage, current, and conductance arrays must have both a real and an imaginary component. During AC analysis, the system matrices become complex quantities.

In addition to storing the DC models, the circuit capacitor and inductor impedances are added to the complex conductance arrays.

Once the complex currents, voltages, and impedances of the circuit elements are stored in the solution arrays, SPICE solves the system equations at each frequency point in the analysis. At each frequency point, the frequency-dependent impedances are calculated and stored in the conductance array. The solution algorithm then determines the complex voltages and currents which satisfy the circuit equations.

The invocation of the linear models simplifies the task of finding the solution to the circuit equations. (Remember, SPICE only uses the non-linear solver when nonlinear elements are present.) In the AC frequency sweep, after the bias point has been found, SPICE uses the computer equivalent of Gaussian elimination (LU decomposition) to solve the linear system equations at each frequency point. Because the arrays contain only linear elements, SPICE solves the small-signal voltages in one Gaussian decomposition (SPICE does not iterate on the AC small-signal solution), and because of this, the AC analysis tends to be much faster than any other sweep analysis.

Transient time sweep analysis

Of all the analysis types, the transient analysis is the most complicated. In many aspects, the transient analysis is similar to the DC sweep analysis but adds several complicating factors to the analysis.

The transient analysis starts as with a DC bias point calculation. During this calculation, the circuit capacitors are modeled as open circuits and the circuit inductors are modeled as short circuits. The bias point of the circuit describes the state of the circuit at time $T=0$ during the transient analysis.

But the instant after the bias point has been found, the analysis changes. For every analysis point after the DC bias point, the time-dependent capacitor and inductor impedances must be added to the system equations. To do this, SPICE uses a numeric integration routine to transform the circuit inductors and capacitors into simplified equivalent circuits. The equivalent circuit of a capacitor or an inductor represents the instantaneous I-V relationship of the device, and the equivalent circuit model changes at every timepoint in the analysis. At each new timepoint, the values from the capacitor and inductor equivalent circuits are stored in the current and conductance arrays.

After the DC bias has been found, SPICE switches into the transient solution mode. The transient solver is similar to the DC sweep solver except with the addition of the capacitor and inductor equivalent circuits. At each timepoint in the analysis, SPICE goes through a series of

Newton-Raphson iterations. As in the DC sweep analysis, the starting point for the transient iterative search is the previous set of node voltages from the last solution timepoint. As the transient analysis proceeds, SPICE uses the previous solution voltages as an initial guess for the next series of iterations. The iterative process continues until the solution voltages for that timepoint are found, or the number of iterations exceeds the allowed limit. As the solution voltages for each timepoint are found, the voltage and current values named on the .PRINT statement are saved in memory. After the transient analysis is complete, SPICE prints the stored solution points in the output in either tabular or graphical form.

Summary

On the surface, circuit simulation is a complex process of numerical analysis. But if you break each of the procedures into simple blocks, the process becomes much more understandable. Any of the analysis results which SPICE computes could also be predicted with simple hand calculations (although it might take years to do the same number of calculations SPICE can do in seconds). There is nothing magical going on in SPICE.

The focus of this text is to show SPICE users how each of the algorithms works, why sometimes they don't work, which type of circuits lead the simulator to highly accurate results, and which circuits lead the simulator to erroneous results. Maybe most important of all, in this text SPICE is seen as a *tool*, a tool with a unique set of capabilities and a unique set of limitations.

Like every tool, SPICE has limitations and does not always produce the desired result. (Using a hammer as a screwdriver rarely leads to the desired result either.) Whether the tool is an automobile engine or a hammer drill or a circuit simulator, to use a tool properly the abilities and limitations of the tool must be understood. From the user's point of view, the limits of software tools are often difficult to define, *because looking inside software is nearly impossible,* and in many cases illegal.

This text examines SPICE and the internal workings of the simulator. Understanding how SPICE works, understanding the abilities and limitations of the simulator, and understanding why and where the simulator introduces inaccuracies are the keys to using SPICE effectively and productively.

References

1. J. Vlach, K. Singhal, *Computer Methods for Circuit Analysis and Design,* Van Nostrand Reinhold, 1983, ISBN 0-442-28108-0.
2. W. Hayt, J. Kemmerly, *Engineering Circuit Analysis,* McGraw-Hill, 1978, ISBN 0-07-027393-6.

3

Nonconvergence

Understanding Nonconvergence

Many SPICE users have, at one time or another, observed a simulation which fails to converge. Nonconvergence is the failure of the nonlinear solution algorithm; it means the simulator failed to find a set of node voltages and branch currents which conform to Kirchhoff's voltage and current laws. Nonconvergence is probably the most persistent and frustrating problem facing simulation users.

But what causes nonconvergence? Is it caused by the solution algorithm or by the type of analysis being performed? Do the models play a role in nonconvergence? And, the most important question of all, what can be done to minimize or eliminate nonconvergence?

Although computer simulation has been with us for over 20 years, and nonconvergence has been with us since the beginning of computer simulation, very little has been written about the causes and cures of nonconvergence. Many experienced simulation users learned to overcome nonconvergence problems with years of experience cultivated from trial and error experimentation. But learning through trial and error is inefficient, ineffective, and very frustrating.

In this chapter, the mechanisms which cause nonconvergence will be explored and explained. Each section of this chapter will examine a different nonconvergence mechanism. Once the nonconvergence mechanisms are known, a simple, methodical, systematic technique will be illustrated to eliminate the nonconvergence catalists.

Convergence and
the Newton-Raphson Algorithm

The study of nonconvergence is best started with a brief discussion of how SPICE converges on a set of node voltages. Understanding how

SPICE converges on a set of node voltages will lead to a better understanding of why SPICE sometimes fails to converge. Chapter 2 demonstrates how the Newton-Raphson algorithm is applied to a set of nonlinear circuit equations, but there is a lot to be learned by looking at the Newton-Raphson process graphically.

Figure 3.1 shows the simple diode circuit used in Chap. 2 to illustrate the Newton-Raphson iterative solution technique. This same circuit will be used to illustrate the Newton-Raphson process graphically.

Figure 3.2a illustrates the I-V characteristics of the circuit diode, and 3.2b illustrates the load line imposed by the resistor and current source. If Figs. 3.2a and b are superimposed, the voltage at node VD may be determined from the intersection of the diode I-V characteristics and the load line as shown in Fig. 3.3. This is a graphical technique for finding the voltage at node VD.

In SPICE, the voltage VD is found with the Newton-Raphson algorithm. Figure 3.4a illustrates the superposition of the diode I-V characteristics and the load line of the diode circuit. The Newton-Raphson

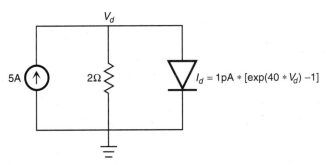

Figure 3.1   A simple diode circuit.

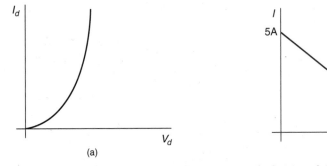

Figure 3.2   The I-V characteristic of Fig. 3.1 circuit diode (a); and the load line of Fig. 3.1 linear elements (b).

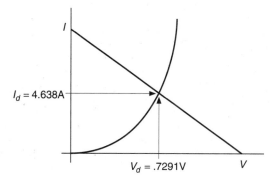

Figure 3.3 The solution of Fig. 3.1 circuit is found at the intersection of the load line and I-V characteristic of the diode.

algorithm begins searching for the intersection of the two curves with an initial voltage guess. After the initial guess (V_0) has been chosen, SPICE passes the V_0 voltage to the LOAD subroutine. The LOAD subroutine determines the linear model of the diode at the voltage V_0. Figure 3.4b illustrates the linear model of the diode superimposed on Fig. 3.4a. Once the linear model has been defined, the Newton-Raphson algorithm calculates the second iterative voltage value (V_1) by locating the intersection of the linear model and the circuit load line as shown in Fig. 3.4c.

The iterative process continues as the linear model of the diode is recomputed for the voltage V_1. The new linear model is again extended to the circuit load line to predict the third iterative voltage value (V_2) as shown in Fig. 3.4d.

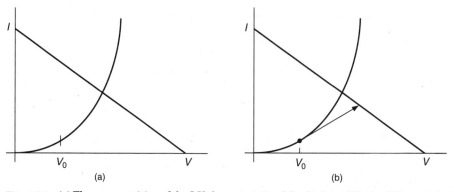

Figure 3.4 (a) The superposition of the I-V characteristic of the diode and the load line and the initial iterative guess V_0; (b) the linear approximation to the diode characteristics at V_0.

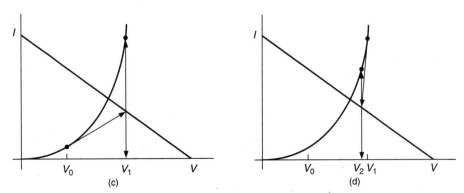

Figure 3.4 (c) The linear approximation to the diode characteristic and the load line determine the iterative voltage V_1; (d) the linear approximation to the diode curve at V_1 and the load line determine V_2.

The Newton-Raphson iterations continue, and eventually the iterative voltage values approach the intersection of the diode current and the circuit load line. As the iterative voltage values approach the exact intersection of the two curves, the difference between sequential voltage iterations decreases, as shown in Fig. 3.5. SPICE converges on the solution voltage and stops iterating when the voltage-to-voltage iterations are within a predefined error tolerance. Figure 3.6 illustrates the iterative node voltage values and the error between iterations. The error tolerance is an important part of the solution algorithm and defines how close SPICE converges to the exact solution before termi-

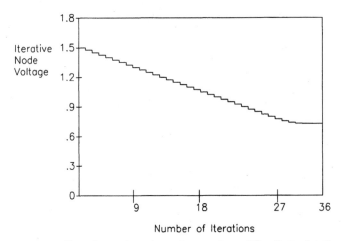

Figure 3.5   Iteration-to-iteration voltage values of the diode circuit with an initial voltage estimate of 1.5 volts. (*Reprinted from* Successfully Simulating Circuits with SPICE. *Used with permission.*)

nating the iterative process. The error tolerances are defined by the user on the .OPTIONS statement.

Causes of Nonconvergence

There are several reasons the simulator may fail to converge. Some of the causes may be traced to the Newton-Raphson algorithm, some to the type of analysis being performed, and others to the device models. Since there are several nonconvergence mechanisms, learning to identify where the nonconvergence came from is an important lesson in overcoming nonconvergence failures.

Overcoming Nonconvergence Failures

Seasoned users will testify to the problems of nonconvergence while newer users may only have seen problems occasionally. For those who have not seen nonconvergence, simulate the circuit file ch3–1.cir with the command:

```
SIM CH3-1.CIR
```

Then look at the output file. In this simulation, SPICE will not be able to determine the proper DC bias of the circuit.

The ch3–1.cir is a typical nonconvergence failure, and the nonconvergence can be corrected with the proper application of .OPTIONS statement parameters. In fact, *somewhere between 75 to 90 percent of*

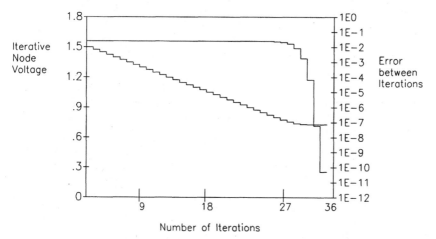

Figure 3.6   Iteration-to-iteration voltage values and the error between iterations for the diode circuit. (*Reprinted from* Successfully Simulating Circuits with SPICE. *Used with permission.*)

*all nonconvergence occurrences may be eliminated with proper applica-
tion of the parameters in the .OPTIONS statement or model parameters
on the .MODEL statement.* Unfortunately for many users, the .OPTIONS statement parameters
have never been clearly defined, nor have the true meaning of the
parameters been explained. Due to the lack of good documentation,
many users simply rely on trying a variety of option settings until one
of them will correct the nonconvergence problem. But this technique is
neither very scientific nor very reliable, and this technique is only
marginally successful.

In this chapter, the causes of nonconvergence will be explained and,
for each of the causes, simple step-by-step procedures will be suggested
to correct the cause of the problem. This chapter focuses on resolving
the problems of nonconvergence.

Problem-solving approach

In addressing the problem of nonconvergence, several different causes
must be examined. Some nonconvergence problems are caused by the
Newton-Raphson algorithm as it is applied in SPICE. These problems
will be addressed in the section titled "General Newton-Raphson Convergence Aids," and the techniques presented here will apply to every
nonlinear analysis in SPICE. The General Newton-Raphson Convergence Aids will reduce nonconvergence problems during the DC operating point analysis, the DC sweep analysis, and the transient time
sweep. The General Newton-Raphson Convergence Aids define the
error tolerance for convergence and set the minimum and maximum
allowed conductance values of the circuit. Many nonconvergence problems will be eliminated simply by setting the General Newton-Raphson
Convergence Aids parameters. Because these convergence aids apply to
all analysis types, the General Newton-Raphson Convergence Aids
should be applied to every simulation before the analysis begins.

But the General Newton-Raphson Convergence Aids will not eliminate all of the nonconvergence problems designers see. Many nonconvergence failures can be linked to the type of analysis being performed.
For this reason, the sections titled "Nonconvergence and the DC Operating Point Solution," "Nonconvergence and the DC Sweep Analysis,"
and "Nonconvergence and the Transient Analysis" will look at the root
causes of nonconvergence which are related to specific analysis types.
Again in each of these sections, the cause of nonconvergence will be
discussed and simple step-by-step procedures will be presented to
eliminate the cause of the nonconvergence.

In essence, this chapter is a cookbook approach to resolving nonconvergence problems. First, set the General Newton-Raphson Conver-

gence Aids parameters. Then, if your simulation fails because of non-convergence, determine which analysis type SPICE was performing when the nonconvergence occurred and follow the procedures outlined for that analysis type.

General Newton-Raphson Convergence Aids

The General Newton-Raphson Convergence Aids are a set of .OPTIONS statement parameters and model parameters which should be set to appropriate levels for your circuit. The General Newton-Raphson Convergence Aids define the accuracy SPICE must achieve to converge on a solution and stop the iterative procedure. The General Newton-Raphson Convergence Aids also determine the minimum and maximum allowed circuit resistance values; these values will determine how quickly SPICE converges to the proper solution.

The acronym SPICE stands for Simulation Program with Integrated Circuit Emphasis. During the late 1960s and early 1970s, breadboarding was found to be nearly impossible for many integrated circuits, and the need to analyze integrated circuit behavior was crucial to the development of a successful design. Much of the funding for the development of SPICE came from the integrated circuits industry, and therefore many of the default values in SPICE were selected for these types of circuits.

But today, in addition to integrated circuits, SPICE is used to simulate discrete, board-level, and even high-power circuits. Many of the program's default settings are inappropriate for these types of circuits. The General Newton-Raphson Convergence Aids reset the program's default settings to levels appropriate for the circuit being simulated.

The General Newton-Raphson Convergence Aids consist of learning to set the accuracy error tolerances (RELTOL, VNTOL, and ABSTOL), learning to set the minimum conductance GMIN, and learning to set the series resistance parameters of the semiconductor models.

Accuracy error tolerances

In the early versions of CANCER and SPICE1, convergence was achieved when the program found a set of node voltages which satisfied Kirchhoff's voltage law or was sufficiently close to those voltages. For diodes, bipolar transistors, and other devices which possess an exponential I-V characteristic, a small change in voltage may produce a large change in device current. This behavior is shown in Fig. 3.7. Neither CANCER nor SPICE1 checked for convergence of the nonlinear device branch currents. Often, simulating circuits which contained exponential I-V characteristics led to results which did not satisfy

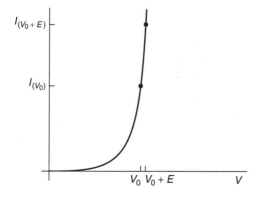

Figure 3.7 For forward-biased PN junctions, a small change in voltage may lead to a large change in current.

Kirchhoff's current law; because of this, later versions of SPICE (including all versions of SPICE2) added a mechanism to check for current convergence in addition to the mechanism which checks for voltage convergence. This means that not only do the voltage-to-voltage iterations have to converge on a solution, the nonlinear device branch currents must converge on a solution as well. This also means that error tolerances for both voltage and current must be defined.

The error tolerances are important for two reasons. First, the error tolerances define the accuracy of the solution. Second, the error tolerances define how many iterations are required to find the solution. Figure 3.8 illustrates how a different error tolerance will require more or fewer iterations to achieve a given level of accuracy.

In SPICE, both the voltage error tolerance and the current error tolerance are composed of a relative limit and an absolute limit. Figure 3.9 illustrates the procedure SPICE uses to check for convergence during the iterative process.

During the iterative process, as SPICE gets closer to the exact set of node voltages which solves the circuit equations, the per-iteration voltage change and the per-iteration current change become very small. Ideally, SPICE would stop iterating when the per-iteration voltage change and the per-iteration current change fell to zero. In this case, the voltage and current values between iterations would be exactly equal. But because of the round-off errors associated with digital computers, SPICE defines convergence when the per-iteration node voltage for every node and the per-iteration current for every nonlinear branch is less than the error tolerance defined by Fig. 3.9.

After every Newton-Raphson iteration, SPICE checks the error between the previous and the present iterative node voltages, and also checks the error between the previous and present iterative branch currents. Once SPICE has found a solution which produces nearly

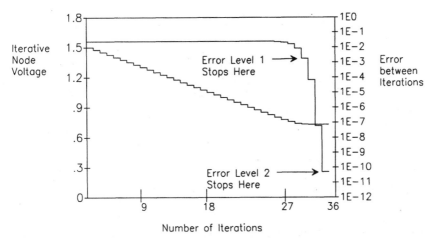

Figure 3.8 To reach higher levels of solution accuracy, more iterations are required. (*Reprinted from* Successfully Simulating Circuits with SPICE. *Used with permission.*)

identical results between two or more iterations, the iterations are stopped, and the solution voltages and currents are saved. If a solution is not found within the allowed number of iterations, SPICE terminates the iterative process, prints the nonconvergence warning message, and halts the simulation.

```
//    NC is the number of nonconvergence circuit nodes
//    or nonlinear branch currents
ITERATION_NUMBER = ITERATION_NUMBER+1
NC = 0
DO I=1, # CIRCUIT NODES
   IF ( |V(n) - V(n-1)| > RELTOL*V(n) + VNTOL ) NC=NC+1
ENDDO
DO J=1, # NONLINEAR BRANCH CURRENTS
   IF ( |I(n) - I(n-1)| > RELTOL*I(n) + ABSTOL ) NC=NC+1
ENDDO
//    If all the nodes and branches have converged, we have a
//    valid solution point. If not, perform one more Newton
//    iteration.
IF (NC = 0)THEN
   GO TO SAVE_SOLUTION_POINT
ELSE
   IF (ITERATION_NUMBER < ITERATION_LIMIT) THEN
      GO TO PERFORM_ANOTHER_NEWTON_ITERATION
   ELSE
   GO TO SOLUTION_FAILED_TO_CONVERGE
   ENDIF
ENDIF
```

Figure 3.9 Checking for a convergent solution voltage.

The relative error tolerance. The error tolerance is defined by the .OPTIONS statement parameters RELTOL, VNTOL, and ABSTOL. RELTOL defines the relative error tolerance for convergence. The relative error tolerance requires the iterative process to continue until the per-iteration voltage change and the per-iteration current change is less than a percentage of the final result. For example, RELTOL is set by default to .001 (.1 percent). For a 5-volt circuit node, the node voltages predicted between iterations must be 5mV or less to satisfy the relative error tolerance convergence criterion. When the node voltage is within 5mV of previous iteration, SPICE will terminate the iterative process. A 5mA branch current must converge to 5uA or less to terminate the iterative process. RELTOL may be reset by the user with the .OPTIONS statement.

The absolute error tolerance. In Fig. 3.9, the relative error tolerance RELTOL is complemented with the absolute error tolerances VNTOL and ABSTOL. The absolute error tolerances are necessary because when a node voltage or branch current approaches or crosses zero, the relative error tolerance approaches zero, which means the simulation result must be infinitely accurate. To avoid problems under these conditions, the authors of SPICE added the absolute error tolerances VNTOL and ABSTOL to the convergence check.

As shown in Fig. 3.9, to achieve convergence, the per-iteration voltage change and the per-iteration current change must be less than the sum of the relative error tolerance and the absolute error tolerance. But VNTOL has a default value of only 1uV and ABSTOL has a default value of only 1pA. Under normal operating conditions, the relative error tolerance is much larger than the absolute error tolerance, and, under these conditions, convergence is defined by the relative error tolerance. But when a node voltage or branch current falls to or approaches zero, the relative error tolerance falls to zero too. Under these conditions the absolute error tolerances define when the iterations stop.

Setting the error tolerances. For many types of circuits, the default setting of RELTOL produces an acceptable amount of accuracy and a reasonable simulation run time. A few circuits will require a more accurate solution (smaller RELTOL), and some may require less accuracy (larger RELTOL). The value of RELTOL can be set on the .OPTION statement as shown below.

```
.OPTIONS RELTOL=.0001
```

Changing the value of RELTOL will change the amount of time a simulation requires. A smaller RELTOL increases the accuracy of the result but takes longer to simulate because of the extra iterations required to achieve the extra accuracy. A larger RELTOL decreases the accuracy of the result but simulates faster. As a general guideline, decreasing RELTOL by a factor of ten (more accurate result) approximately doubles the number of iterations required to solve the circuit. Increasing RELTOL by a factor of ten (less accurate result) approximately halves the number of iterations required to solve the circuit.

The default values for RELTOL, VNTOL, and ABSTOL were set to produce an acceptable result for the type of integrated circuits being simulated in the late 1970s. (Remember SPICE is an acronym for Simulation Program with Integrated Circuit Emphasis.) But today, SPICE is used to simulate many different types of circuits including discrete, board-level, and high-power circuits. While RELTOL produces the same percentage of accuracy regardless of the voltage and current levels in a circuit, circuits which contain voltage and current levels much higher or much lower than typical integrated circuits necessitate resetting VNTOL and ABSTOL to appropriate levels. (Do you really need to simulate your 20-volt switching power supply to an accuracy of 1uV and 1pA?)

The first of the General Newton-Raphson Convergence Aids is proper selection of the simulation error tolerances. To set the error tolerances, follow the procedure outlined in Fig. 3.10. Setting the error tolerances requires knowledge of the voltage and current levels of the circuit, but setting the error tolerances to levels appropriate to your circuit will result in faster simulations and fewer nonconvergence failures.

Once RELTOL has been determined,

set VNTOL = RELTOL * V_{small}

Where V_{small} is the smallest voltage (magnitude) of interest in the circuit.

(For example, V_{small} might be an input offset voltage for an op-amp circuit or a low logic level for a digital circuit.)

Once VNTOL has been determined,

set ABSTOL = RELTOL * I_{small}

Where I_{small} is the smallest current (magnitude) of interest in the circuit.

(For example, I_{small} might be an input offset current for an op-amp circuit or the reverse leakage current of a diode circuit.)

Then, enter the values on the .OPTION statement (.OPTIONS RELTOL=.0001 VNTOL=.0001V ABSTOL=10NA)

Figure 3.10   Setting the SPICE error tolerances.

Circuit conductance values

Like the error tolerances, the conductance values of the circuit also determine how quickly SPICE converges on a solution. But too few SPICE users realize the importance of the conductance terms. When SPICE begins the iterative process, the solution voltages and currents must be found within an allowed number of iterations, and, although the maximum and minimum conductances have a minor impact on the accuracy of the circuit, the upper and lower limits on conductance will influence how quickly the Newton-Raphson algorithm converges to a solution.

$$V_{n+1} = V_n - \frac{F(V_n)}{G(V_n)} \tag{3.1}$$

Equation 3.1 is the equation of the Newton-Raphson algorithm when applied to a single node circuit. In Eq. 3.1, the denominator of the second term is the conductance of the circuit elements. The Newton-Raphson algorithm uses conductance values to predict where the next voltage iteration will fall. If the conductance becomes very small, the second term of Eq. 3.1 ($F(V_n)/G(V_n)$) becomes quite large. Worse yet, if the conductance value ever reaches zero, the next Newton iteration will cause a floating divide-by-zero error and crash the program.

Figure 3.11 illustrates the diode characteristics under reverse bias conditions. In this region, SPICE models the I-V characteristics as a constant current equal to –IS. When the current is constant and no longer a function of the diode voltage, the conductance of the diode goes to zero. Under these conditions, the Newton-Raphson algorithm should fail because the diode conductance is zero.

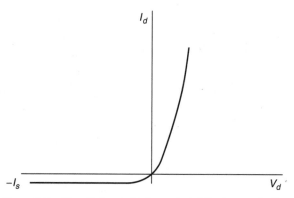

Figure 3.11   The diode conductance (g_d) falls to zero in the reverse-bias region of operation.

Since all of the semiconductor devices contain one or more regions of zero conductance (constant current output), the conductance problem had to be resolved. The authors of SPICE resolved this problem by placing a shunt resistor in parallel with every PN junction of every semiconductor model in SPICE. The resistor has a default value of 1000G ohms, and the value of the resistor is set with the .OPTION GMIN=X. GMIN represents the conductance (reciprocal of the resistance) of the shunt resistor. The GMIN resistor is built into the model equations and always provides a small voltage-dependent current for the devices. Equation 3.2 is the diode equation under reverse bias with the GMIN resistor term included, and Eq. 3.3 is the conductance of the diode.

$$I_d = -IS + V_d*\text{GMIN} \qquad (3.2)$$

$$G_d = \frac{dI_d}{dV_d} = 0 + \text{GMIN} \qquad (3.3)$$

The GMIN resistor is found across the diode, the base-emitter and base-collector junctions of the bipolar, the gate-drain and gate-source junctions of the JFET, and the drain-bulk and source-bulk junctions of the MOSFET. Every PN junction of every device in SPICE includes the GMIN resistor. Normally the resistance value is so high that the resulting current is significantly below the error tolerances of the simulation. This means the current through the GMIN resistor does not contribute to the value of the simulation result. For example, if a circuit diode is held in at −5V reverse bias with 1uA reverse saturation current, the current flowing through the GMIN resistor is 5pA (5V/1000G ohms = 5pA). 5pA is less than the 1nA relative error (RELTOL) tolerance of SPICE (1uA * .001 = 1nA). The default value of GMIN was chosen so the simulation accuracy would not be affected by its presence.

But how does GMIN influence the convergence characteristics of the circuit? As long as the conductance term in Eq. 3.1 is finite, the Newton-Raphson algorithm will continue iterating to a solution, but if the conductance is very small, the second term of Eq. 3.1 ($F(V_n)/(V_n)$) becomes quite large, and the next iterative voltage value will be far from the previous voltage value as illustrated in Fig. 3.12. Often, very small conductance values force the Newton-Raphson algorithm to significantly overshoot the correct solution voltage. When this occurs, the Newton-Raphson algorithm will require dozens of iterations to work back to the correct solution voltage.

This effect can be demonstrated with the use of the DIODE.EXE program found in the ch2 subdirectory of the text floppy disk. Starting with an initial diode voltage of .5 volts, the first iteration is 9.6 volts!

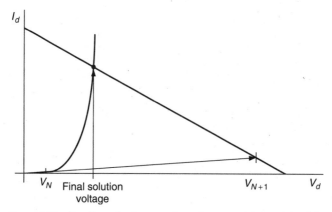

Figure 3.12 Small conductance values (large resistances) lead to large iteration-to-iteration voltage changes.

To reach the final solution voltage of .7291 requires an additional 360 iterations. With an initial diode voltage of .55 volts, the first iteration is 7.8 volts and requires 289 additional iterations to reach .7291 volts. An initial voltage of .6 volts generates 3.6 volts on the first iteration and requires 118 more iterations to reach .7291 volts.

SPICE must find the proper solution value within a fixed number of iterations. Because of this restriction, finding the solution with as few iterations as possible is an important aspect of achieving convergence. For simulation users, the larger the value of GMIN, the faster the Newton-Raphson algorithm will converge on a solution. *To help avoid nonconvergence problems, GMIN should be set as large as possible without affecting the accuracy of the simulation output.* Raising GMIN decreases the size of the shunt resistor. As long as the shunt resistor's current contribution is lower than the relative error tolerance current resolution, there will be no difference in the accuracy of the simulation result.

Figure 3.13 illustrates how to select a value of GMIN that is appropriate for your circuit.

GMIN is a global setting for the entire circuit, so when selecting the value of GMIN, consider the most current sensitive portions of the circuit.

Just as the Newton-Raphson algorithm has problems with very small values of conductance, very large values of conductance can lead to problems also. Figure 3.14 illustrates the I-V characteristics of a diode under high forward bias conditions. For very large values of conductance, the second term of Eq. 3.1 becomes very small. Because of this, the next voltage iteration (V_{n+1}) will be only slightly different than V_n, and if the device is biased far from the proper solution voltage, many iterations will be required to work back to the proper solution.

To Set GMIN

Determine the smallest parasitic resistance value (*Rp*) which could be placed across any two nodes without influencing the behavior of the circuit.

Set `GMIN = 1/Rp`

Enter the value of GMIN on the .OPTIONS statement (`.OPTIONS GMIN=1E-10`)

Figure 3.13   Procedure to set the GMIN value.

High conductance values are extremely troublesome when diodes or PN junctions have a forward bias of more than .8 volts. Beyond this point, the conductance values become unrealistically large (resistance becomes very small) and will lead to nonconvergence problems. While very few circuits require diode biases this high, during the iterative process, SPICE may apply a high forward bias to any of the diodes in the circuit while iterating to a solution. If this occurs, SPICE may fail to converge before reaching the proper circuit solution.

But SPICE users can guard against the extremely large values of conductance seen in diodes and PN junctions by always specifying the series resistance model parameter for all circuit diodes and bipolar devices. (While it is a good habit to always set the series resistance for the MOSFET and JFET devices, these are less prone to forward biasing the internal PN junctions of the device.) The default value of series resistance is zero ohms, and if during the iterative process SPICE applies a high forward bias voltage to a diode or PN junction, a nonconvergence condition may occur. But if the series resistance terms nonzero, under high forward bias conditions, the series resistance

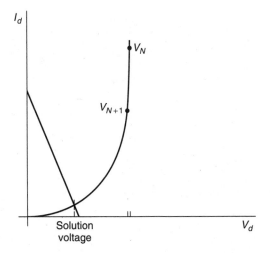

Figure 3.14   Large conductance values (small resistance values) lead to small iteration-to-iteration voltage changes.

dominates the conductance of the device and helps reduce the occurrence of nonconvergence.

The series resistance terms are found on the .MODEL parameter statements for each of the semiconductor devices and are not global parameters found on the .OPTION statement. Look at the model parameters of your circuit. A good model should always have the series resistance defined. If the model does not have a resistance term in the parameter list, select a value of series resistance which is small enough to remain unobtrusive to the operation of the circuit and add it to the model parameters. Table 3.1 illustrates the series resistance terms for each of the semiconductor devices.

General Newton-Raphson Convergence Aids Summary

The General Newton-Raphson Convergence Aids include learning to set the error tolerances for the circuit, choosing a value of GMIN which reflects the largest resistances in the circuit, and setting the series resistance terms for diodes and bipolar transistors to a nonzero value. By following these guidelines, simulations will run faster and fail to converge less often. Table 3.2 illustrates the General Newton-Raphson Convergence Aids.

Analysis-Specific Convergence Aids

Specifying the General Newton-Raphson Convergence Aids will reduce the occurrence of nonconvergence. But many nonconvergence failures are tied to one or more analysis types. In many of these cases, the General Newton-Raphson Convergence Aids will not be enough to eliminate the nonconvergence. In most cases of an analysis-specific nonconvergence failure, the nonconvergence may be corrected by understanding what caused the problem and setting the appropriate .OPTION statements. The key to overcoming analysis-specific nonconvergence failures is to understand what caused the failure, and understanding the cause of nonconvergence is where most engineers have trouble.

SPICE is capable of many different analysis types, but only three of these may fail from nonconvergence. SPICE may fail to converge dur-

TABLE 3.1 Series Resistance Model Parameters

Semiconductor type	Series resistance parameters
Diode	RS
Bipolar transistor	RE and RC
JFET	RD and RS
MOSFET	RD and RS

TABLE 3.2   General Newton-Raphson Convergence Aids

Convergence aid	Type of circuits
Error tolerances (.OPTION RELTOL=X +VNTOL=X ABSTOL=X)	Discrete circuits Hybrid circuits Board-level circuits Power circuits Power-integrated circuits
GMIN (.OPTIONS GMIN=X)	All circuits
Series resistance model parameters (.MODEL DMOD D RS=X .MODEL QMOD NPN RE=X RC=X)	Diode and bipolar circuits

ing the DC operating point calculation, during a DC sweep analysis, or during a transient time sweep. In overcoming analysis-specific non-convergence, the sequence of events SPICE follows for a given analysis must be known. Table 3.3 illustrates the analysis flow for the four major SPICE analysis types.

Every analysis type in SPICE starts with a DC bias point calculation. For this reason, the authors of SPICE added more DC bias point non-convergence aids than any other analysis-specific nonconvergence aids.

DC Bias Point Convergence Aids

The task of determining the bias point of a circuit is both the most difficult and the most important. The bias point is important because it serves at the initial circuit condition for all other analysis types. The bias point calculation is difficult because often SPICE has little or no information on how the circuit should be biased. (Compare this with the DC sweep analysis or transient time sweep where the prior solution voltages are used as the initial guess for the next series of Newton iterations.)

TABLE 3.3   Analysis Flow in SPICE

Analysis type	Analysis flow
DC operating point	DC bias calculation
DC sweep analysis	DC bias calculation DC sweep calculation (Use bias point as first point in DC sweep.)
AC frequency sweep	DC bias calculation AC frequency sweep (Use the bias point to determine the linear small-signal models of all nonlinear devices.)
Transient time sweep	DC bias calculation Transient time sweep (Use the bias point to determine the time $T=0$ value of capacitor voltage and inductor current.)

Functionally, the DC bias point calculation has already been covered both in the diode circuit example earlier in this chapter and in Chap. 2. But the specifics of how the DC bias calculation is performed and the impact of the .NODESET statement and the ITL1 option have yet to be shown.

DC bias point calculation

As shown in Chap. 2, SPICE constructs the system (matrix) equations from the elements found in the input file. Once the system equations are in place, SPICE needs an initial guess for the circuit node voltages to start the Newton-Raphson algorithm.

SPICE makes the initial node voltage array guess by setting any nodes connected to a voltage source to the time zero or DC level described in the input file. The nodes in the current array which represent connects to a circuit current source are set to the appropriate time zero or DC current level. All the remaining nodes in the circuit are set to zero, and this forms the initial guess to start the Newton-Raphson algorithm.

Once the initial guess is entered in the voltage array, SPICE begins the iterative process. The Newton iterations continue until either all of the nodes and all of the nonlinear branch currents converge to within the specified error tolerances or the number of iterations exceeds ITL1. ITL1 is the .OPTIONS statement parameter which determines the maximum number of iterations SPICE can use to determine the DC operating point of a circuit. If all the nodes and branch currents have not converged to within the specified error tolerances before ITL1 iterations have been used, SPICE aborts the iterative process and prints the dreaded "NO CONVERGENCE IN DC OPERATING POINT" error message. ITL1 defaults to 100 in SPICE.

If SPICE fails to converge to a DC bias solution, the cause can most often be traced to either lack of a good initial guess to start the Newton-Raphson iterations, or the lack of enough iterations to reach a satisfactory solution for the specified error tolerances.

Raising ITL1

The first DC bias point nonconvergence aid is ITL1. Many cases of nonconvergence can be eliminated by simply raising ITL1 on the .OPTIONS statement. The default setting of ITL1 was determined in the mid-1970s when simulating more than a few dozen nodes was reason for a coffee break. Today circuits with hundreds or thousands of nodes are common. Many circuits will require more than 100 iterations to reach a stable bias point. *The first rule of the DC bias point noncon-*

vergence aids is to raise ITL1. ITL1 sets the limit on the number of iterations SPICE uses before the search for the bias point is aborted. By raising ITL1, the extra iterations will only be used by SPICE if they are needed. The .OPTIONS statement to raise ITL1 is shown here.

```
.OPTIONS ITL1=500
```

Table 3.4 shows the number of iterations and the percentage of circuits which converged to a stable DC bias point with no other convergence aids added.[1]

From Table 3.4, setting ITL1 to 500 yielded the best results. In the circuits tested, no measurable improvement was observed by setting ITL1 higher. For all circuits, *set ITL1 to 500 on the .OPTIONS statement.* Some circuits may require more, most less. ITL1 is the first of the DC operating point nonconvergence aids to set.

As an example, simulate the ch3-15a.cir circuit file with the command:

```
SIM CH3-15A.CIR
```

After 100 iterations, the simulation will terminate with the warning "DC Operating Point Nonconvergence." Then simulate ch3-15b.cir. The ch3-15a and ch3-15b circuits are identical except for the .OPTIONS ITL1=500 statement. The ch3-15b circuit allows SPICE to use the extra iterations required to find the correct circuit bias.

Setting initial node voltages

The DIODE.EXE program from Chap. 2 illustrates the importance of starting the Newton-Raphson algorithm with an initial guess which is close to the final value. Setting the initial node voltages reduces the number of Newton-Raphson iterations required to reach a solution. In SPICE, circuit nodes can be set to an initial value when starting an analysis with the .NODESET statement. Below is an example of using the .NODESET statement to set several circuit node voltages.

```
.NODESET V(2)=12.5 V(5)=5.0 V(7)=16.8
```

TABLE 3.4

Number of iterations to converge	Percentage of circuits which converged
100	60
200	75
500	92
1000	92

The .NODESET statement forces the designated nodes to a specific voltage by applying a Norton-equivalent voltage source to the specified node. The Norton-equivalent source is shown in Fig. 3.15. The source contains a 1-ohm resistor and a current source. The current source's current is set to the voltage specified on the .NODESET statement for that node. For most circuits, the load seen at the specified node is much larger than 1 ohm. Under these conditions, all of the current from the generator is forced through the resistor, thereby forcing the node to the proper voltage.

When SPICE encounters one or more .NODESET statements, the simulator performs two DC bias point calculations. The first calculation is done while the Norton-equivalent sources are held in the circuit. If the circuit converges with the Norton-equivalent sources in place, the bias point voltages are saved, the Norton-equivalent sources are removed, and a second series of Newton-Raphson iterations determines the final bias solution. The second bias point calculation is done to ensure the Norton-equivalent sources do not impose a load on the circuit.

Circuits of moderate complexity with a single, stable bias point rarely require the use of the .NODESET statement. Often, very large circuits will converge faster if several circuit nodes are ".NODESET" to a specified value. But the best use of the .NODESET statement is on circuits with more than one stable operating point. For example, flip-flops and latches may start in either the high or low state. Circuits with hysteresis may start in one of two states. For these types of circuits, the .NODESET statement can be used to enhance convergence and ensure the circuit starts in the proper state.

Often in cases of DC bias point nonconvergence, the number of nodes which fail to converge is small, usually one or two. When SPICE aborts the DC bias calculation, the last iterative node voltage values will be printed in the output file. Inspecting these node values often uncovers

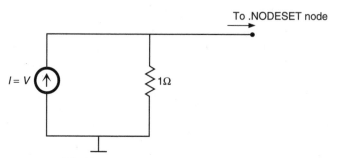

Figure 3.15   The Norton-equivalent voltage source used in SPICE to model .NODESET voltages.

just one or two nodes which are far from the proper bias values. These are the nodes causing the nonconvergence problem, and setting these nodes with the .NODESET statement often results in successfully locating the bias point in the next simulation.

A word of caution about the .NODESET statement: Do not try to apply the .NODESET statement to circuits which do not have a stable DC bias point (such as oscillators and other unstable circuits). When the .NODESET statement is applied to unstable circuits, although SPICE may converge while the Norton-equivalent source is attached to the circuit, during the second series of iterations, after the source has been removed, the instabilities of the circuit will dislocate the stable condition which existed and cause the simulator to fail to converge. SPICE does have a mechanism to start unstable circuits, the .IC statement. But discussion of the .IC statement will be delayed until the section on transient analysis.

Source stepping

In most cases, DC operating point nonconvergence can be eliminated by increasing ITL1 and proper use of the .NODESET statement. But some circuits will prove resistant to even these techniques, and some circuits will contain subcircuits. Nodes within subcircuits cannot be set with a .NODESET statement. For these types of circuits, the Source Stepping algorithm may be used to compute the DC bias.

Source Stepping is a technique which steps the circuit power sources from zero (where the solution to all the nodes is zero) to full power. At each step, the previous node voltages are used as the initial guess for the next series of Newton-Raphson iterations.

The Source Stepping algorithm is used to calculate the DC bias point when the .OPTION ITL6=X is set to a nonzero value. ITL6 is the parameter which determines the number of iterations allowed at each step in the progression. ITL6 is to Source Stepping as ITL1 is to the normal DC bias point calculation. For a circuit which will not converge after raising ITL1 and/or after adding the .NODESET statement, set ITL6 to 500 or more on the .OPTIONS statement. When ITL6 is set to a nonzero value, the Source Stepping algorithm replaces the normal DC bias point calculation.

To demonstrate the Source Stepping algorithm, simulate the ch3-18a.cir disk file with the command:

```
SIM CH3-18A.CIR
```

This large circuit will fail to converge to a DC bias point. Due to the size and complexity of the circuit, appropriate .NODESET values cannot be determined quickly. By replacing the .OPTIONS ITL1=500 with

.OPTIONS ITL6=500 statement, the circuit bias point is quickly located with the Source Stepping algorithm. Simulate the corrected circuit with the command:

```
SIM CH3-18B.CIR
```

At this point, readers might be tempted to use the Source Stepping algorithm in place of the normal DC bias point algorithm for all circuits. While the Source Stepping algorithm works on many nonconvergent circuits, a flaw within the SPICE implementation of the algorithm prevents the Source Stepping method from converging on other circuits.

The Source Stepping algorithm actually begins by ramping the circuit sources down from full power in a binary descending order (full power, ½ power, ¼ power, ⅛ power, 1/16 power). During the ramp-down phase, SPICE is searching for a convergent bias point to serve as the starting point of the ramp-up phase. When a bias point has been found, SPICE ramps the sources up in the same binary order. If, during the ramp-up phase, a nonconvergence occurs, the algorithm repeats the ramp-down, ramp-up phase to work through the nonconvergence. But because of a flaw in the algorithm, during a second ramp-down phase, the power will be held at a local minimum value and will not be ramped below this point. Unable to ramp the sources down further, SPICE quickly exceeds the allowed number of iterations for that step and aborts the DC bias point calculation.

For most circuits, raising ITL1 and using the .NODESET statement properly will produce an accurate bias point solution. When these techniques fail, the Source Stepping algorithm is a good alternate solution technique.

Turning active elements off

SPICE contains one final DC bias point convergence aid known as the OFF statement. The OFF statement, like the .NODESET statement, forces the simulator to compute two DC bias point calculations. During the first calculation, one or more active devices may be turned off. During the second calculation, the active devices are turned on, and the bias point from the previous calculation is used as the initial guess for the second set of Newton iterations.

The OFF statement is added to one or more semiconductor devices in the circuit. Figure 3.16 illustrates the use of the OFF statement for the four semiconductor types in SPICE.

A note of caution about the OFF statement: For many nonlinear circuits, the circuit bias point with one or more active devices turned off

Semiconductor	Netlist statement
Diode	Dxxx na nc D_model OFF
Bipolar transistor	Qxxx nc nb ne Q_model OFF
JFET transistor	Jxxx nd ng ns J_model OFF
MOSFET transistor	Mxxx nd ng ns nb M_model OFF

Figure 3.16   OFF statement syntax for SPICE semiconductor devices.

is substantially different from the bias point when the devices are turned on. For this reason, *the OFF statement rarely achieves convergence on a circuit where the ITL1, the .NODESET statement, and the ITL6 techniques have failed.*

DC bias point convergence aids summary

DC bias point nonconvergence problems can be significantly reduced by following the procedures outlined in this section. Finding the DC bias point is crucial to simulation because the bias point calculation serves as the starting condition for every other analysis in SPICE. For this reason, learning to overcome DC bias point nonconvergence is paramount in producing accurate, high-quality simulation results. Table 3.5 lists the DC bias point convergence aids. Follow the convergence aids in the order shown in Table 3.5 if nonconvergence persists after setting the General Newton-Raphson Convergence Aids.

DC Sweep Convergence Aids

In many ways, performing the DC sweep analysis is like performing a series of DC bias point calculations. To begin the analysis, SPICE performs a DC bias point calculation. Once the bias point has been found,

TABLE 3.5   DC Bias Point Convergence Aids

Convergence aid	Order
Raise ITL1	First
(.OPTION ITL1=500 or more)	
Set .NODESET statements	Second
(.NODESET V(3)=2.1 V(4)=7.9)	
Use Source Stepping	Third
(.OPTION ITL6=500)	
Use OFF statement	Fourth
(Dxxx n+ n- d_model OFF)	

the node voltages are saved and used as the initial guess for the Newton iterations at the next point in the analysis. At each solution point in the analysis, SPICE uses the previous solution as the initial guess for the next series of iterations. This process continues for every point in the sweep analysis. When the analysis is complete, SPICE prints the output result.

But the nonconvergence problems associated with a DC sweep analysis are surprisingly different than the problems of the DC bias point. The two nonconvergence mechanisms of the DC sweep analysis are rapid voltage transitions and model discontinuities. Either of these might cause the simulator to fail to find a solution.

Model discontinuities

The semiconductor device models SPICE uses for simulation were patterned after the physical behavior of real devices. Figure 3.17 illustrates the family of curves from an MOS transistor. Classical physics splits the transistor curves into the linear region and the saturated region of operation. The device equations in SPICE follow these same regions of operation. But, unlike the real device, SPICE uses separate equations for each region of operation. Because of the mathematical difficulty in writing an equation which describes the entire family of curves, two different sets of equations were written, one for the linear region and one for the saturated region, and joined together. Unfortunately, because of the way the equations were joined, there is a discontinuity in the conductance characteristics (the slope of the I-V curve) of the device.[2] In Fig. 3.18, the discontinuity exists at the intersection of the linear and saturation regions.

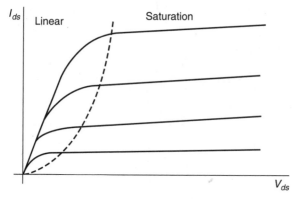

Figure 3.17 The linear and saturated regions of operation of an MOS field-effect transistor.

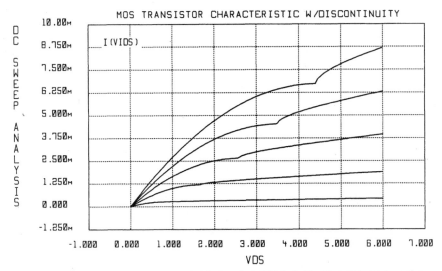

Figure 3.18 Model discontinuities of the Level 3 MOS field-effect SPICE transistor model.

A discontinuity in conductance may lead to problems for the Newton-Raphson algorithm. Figure 3.19*a* illustrates the conductance vs. voltage characteristics around one of the model discontinuities. On the first Newton iteration close to the discontinuity, the conductance value leads to a new iterative voltage value on the other side of the discontinuity. In Fig. 3.19*b*, the next Newton iteration falls on a conductance value which predicts a solution back on the original side of the discontinuity. In Fig. 3.19*c*, the third Newton iteration again predicts a solution on the far side of the discontinuity. When SPICE steps close to or on top of a model discontinuity, the Newton-Raphson iterations may begin to oscillate around the discontinuity. These oscillations use up iterations without progressing towards a solution.

Although model discontinuities may cause a problem when searching for the DC bias point solution, model discontinuity nonconvergence becomes more of a problem during a sweep analysis. While performing the DC bias point calculation, only a single solution point is sought. During the DC sweep analysis, many solution points are required and, in most analyses, the nonlinear circuit elements are being swept through one or more regions of operation. Model discontinuities only pose a problem if the solution steps align with, or fall very close to, the discontinuity. The more steps taken during the sweep analysis, the better the change of stepping into or close to a model discontinuity. This is why model discontinuity nonconvergence is more of a problem during a sweep analysis than during the DC bias point calculation.

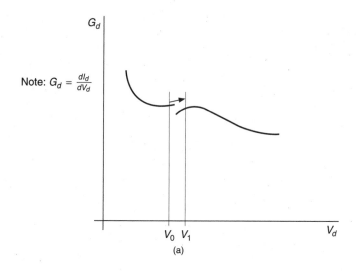

Note: $G_d = \frac{dI_d}{dV_d}$

(a)

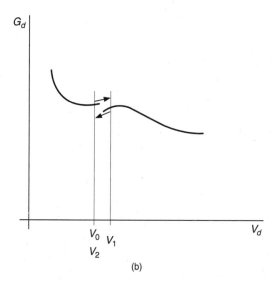

(b)

Figure 3.19(a) and (b) Iterations around a model discontinuity may jump from one side of the discontinuity to the other.

Rapid voltage transitions

The second nonconvergence mechanism found during a DC sweep analysis is caused by rapid voltage or current transitions. Figure 3.20a illustrates the theoretical transfer characteristics (V_{out} vs. V_{in}) of an inverter. The transfer characteristics may be found by performing a DC sweep analysis, sweeping the input from 0 to 5 volts, and observing the output voltage.

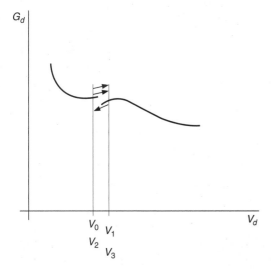

Figure 3.19(c) Model discontinuities may cause oscillation during the iterative process. Such oscillation will cause SPICE to use iterations without approaching a valid solution.

To start the analysis, SPICE determines the DC bias point of the circuit. Once the bias point has been found, the input voltage is raised to the first voltage step of the .DC statement, and the bias point voltages are used as the initial guess for the next series of Newton iterations. *In any SPICE sweep type analysis, the previous solution voltages are used as the initial guess for the next set of Newton iterations.* Using the bias

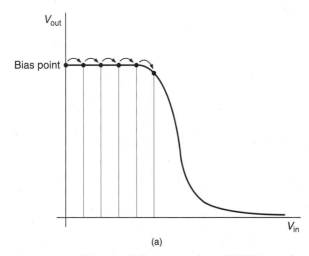

(a)

Figure 3.20(a) During a DC sweep analysis, SPICE uses the previous voltage solution as an initial guess for the next set of Newton iterations.

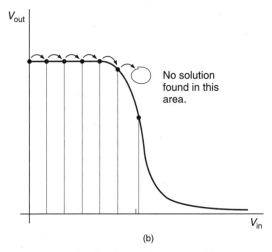

(b)

Figure 3.20(b)   During large voltage transitions, a new solution point may be far from the previous solution point. Under these conditions, SPICE often fails to converge.

voltages as the initial guess, SPICE iterates until a solution for the second sweep point has been found or until ITL2 iterations have been used. ITL2 is the number of iterations allowed at each step in the DC sweep analysis.

This process repeats at each solution point in the sweep analysis. But during large voltage transitions, the voltage change between sweep points may be too large for the Newton iterations to find a solution. Nonconvergence at a transition point is a common DC sweep failure.

When nonconvergence occurs at a voltage transition, many engineers reduce the step size of the .DC statement (Fig. 3.20c). While reducing the step size of the .DC statement will reduce the size of the voltage transition between steps and usually result in a successful simulation, reducing the step size is not the optimum way to eliminate DC sweep voltage-transition nonconvergence failures.

By reducing the .DC statement step increment, many more solution points will be required. Extra solution points require extra iterations and lead to much longer simulation runs. And as shown in Fig. 3.21a and b, reducing the step size increases the chance of stepping into or very close to a model discontinuity and failing to converge because of the resulting oscillation.

Rapid voltage-transition discontinuities are best overcome by raising the number of iterations allowed at each step in the analysis with ITL2. ITL2 limits the number of iterations allowed at each solution point in a DC sweep analysis. By default, ITL2 is set to 50. For many

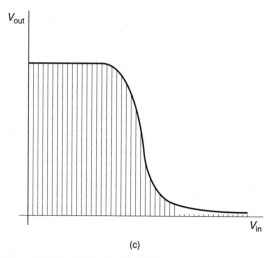

(c)

Figure 3.20(c) Reducing the DC sweep step size is one way of resolving voltage-transition nonconvergence problems. But, reducing the step size increases simulation time and increases the chance of encountering a model discontinuity. Raising the DC sweep iteration limit (ITL2) is the optimum way of resolving voltage-transition nonconvergence problems.

circuits, more than 50 iterations are required to work through large rapid voltage transitions. During rapid voltage transitions, 100, 200, or more iterations may be required to locate the solution. During a DC sweep analysis, for circuits which contain one or more steep transitions, raise ITL2 to 200 on the .OPTIONS statement (.OPTIONS ITL2=200).

If, during a DC sweep analysis, raising ITL2 does not correct the nonconvergence problem, the nonconvergence is probably being caused by one or more model discontinuities. In this case, users have a choice of either obtaining a new model with parameter values which minimize the discontinuity, or increasing or offsetting the analysis step size. Often, model discontinuities may be overcome by increasing or offsetting the analysis step size so the steps are far enough away from the discontinuity to avoid oscillation. Figure 3.22 illustrates how users may increase or offset the DC sweep analysis step size.

DC sweep convergence summary

DC sweep analysis nonconvergence failures are usually caused by one of two failure mechanisms, rapid voltage transitions or device model discontinuities. Overcoming these problems is summarized in Table 3.6.

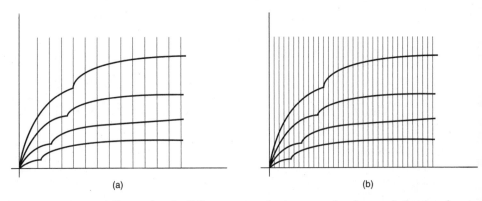

(a)   (b)

Figure 3.21(a) and (b)   Decreasing the DC sweep step size increases the chance of stepping close to a model discontinuity and failing to converge. The smaller the step made, the larger the chance of encountering a discontinuity.

AC Frequency Sweep Convergence Aids

The AC frequency sweep analysis is less prone to nonconvergence problems than any other analysis type in SPICE. The reason for this (as explained in Chap. 2) is that the AC frequency sweep is a linear small-signal analysis which does not include any nonlinear behavior after the bias point has been found. SPICE first determines the DC bias point of the circuit; during the bias-point calculation, SPICE may fail to converge. In these cases, the DC bias point convergence aids may be employed to achieve convergence. After the bias point has been found, the nonlinear device models are all replaced with their linear small-signal equivalents. With no nonlinear devices in the circuit, SPICE switches to the simpler LU decomposition (Gaussian-elimination) solution algorithm. With the LU decomposition solver, no Newton iterations are required to solve the circuit and, because of this, nonconvergence disappears. Once the AC frequency sweep analysis has determined the DC bias point of the circuit, the frequency sweep analysis will always converge. The only nonconvergence problems associated with the AC frequency sweep are problems which arise in finding the DC bias point of the circuit, and these problems have already been discussed in the section titled "DC Bias Point Convergence Aids."

Original	.DC VIN 0V 5V .1V
Increase step	.DC VIN 0V 5V .2V
Offset step	.DC VIN 0.01V 5.01V .1V

Figure 3.22   Changing the analysis step size to overcome nonconvergence.

TABLE 3.6   DC Sweep Convergence Aids

Convergence aid	Cause/order
Raise ITL1 (.OPTIONS ITL2=200)	Rapid voltage transitions/First
Increase or offset analysis step size (See Fig. 3.24.)	Model discontinuity/First
Develop new model parameter set	Model discontinuity/Second

Transient Convergence Aids

For many SPICE users, transient analysis is the analysis type most often used. In many aspects, transient nonconvergence problems are similar to the DC sweep analysis nonconvergence problems, but in some ways the two are quite different.

To begin the transient analysis, SPICE open-circuits the circuit capacitors, short-circuits the circuit inductors, and computes the DC bias point of the circuit. Immediately after the bias point has been found, the capacitors and inductors are restored to the circuit and immediately assume the voltages established by the DC bias point calculation (no charge-up time is required). After the charge-storage elements are placed in the circuit, the first timepoint is calculated, and a series of Newton iterations begins computing the solution voltages and currents at the first timepoint of the analysis. At each new solution point (timepoint), a new set of Newton iterations is used to compute the node voltages and branch currents of the circuit. As the solution algorithm iterates, the numeric integration algorithms determine the capacitor currents and inductor voltages as a function of time. All of this occurs at each timepoint after the DC bias of the circuit has been found.

Dynamic timestep control

Like the DC sweep analysis, nonconvergence problems primarily arise from rapid voltage transitions and device model discontinuities. But, unlike the DC sweep analysis, SPICE uses a dynamic timestep control algorithm to compute the solution points rather than being specified by the user. This comes as a surprise to many users. The analysis statement

```
.TRAN 1NS 100NS
```

is an instruction which directs SPICE to perform a transient analysis from time $T=0$ to $T=100$nS. But the first parameter of the statement, 1nS, is known as the *print interval* and is an instruction to print the output results every 1nS of the analysis. *SPICE does not solve the circuit every 1nS of the analysis.* The internal timestep control algorithm has the job of selecting the actual solution points for the analysis.

During transient analysis, SPICE saves all the solution points in memory. After the analysis is complete, SPICE uses a linear interpolation routine to compute the evenly spaced output points. Figure 3.23 illustrates the difference between the internal (solution) timepoints and the printed results. (The RSPICE option NOINTR forces RSPICE to print these internal timepoints rather than the interpolated timepoints.)

The dynamic timestep control algorithm changes the size of the timestep depending on the circuit activity. During periods of high circuit activity, and large voltage and current changes, the timestep control algorithm maintains a small timestep to help increase accuracy and reduce nonconvergence due to rapid voltage transitions. During periods of low activity, or little or no voltage and current change, the timestep control algorithm increases the step size to speed the simulation to a conclusion.

The dynamic timestep control algorithm also enables the transient analysis to do something no other analysis can do. If SPICE ever fails to converge at a timepoint, the timestep control algorithm automatically discards the nonconvergent timepoint, reduces the timestep to ⅛ of the original size, and reattempts the solution with a smaller timestep. If SPICE fails a second time, the timestep is cut by a factor of eight again, and again, and again, until either the circuit converges or the timestep is reduced below the allowed minimum for the circuit. (The minimum internal timestep is fixed internally and cannot be adjusted by the user.) If the internal timestep is reduced below the allowed minimum, SPICE prints the "Internal Timestep Too Small"

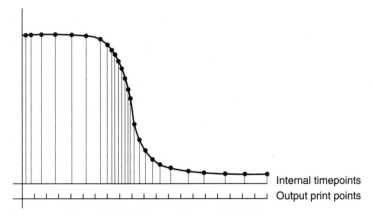

Internal timepoints

Output print points

Figure 3.23 The timepoints SPICE uses during transient analysis are not the same as the points printed in the output file. Many simulators offer an option so users may view the actual timepoints SPICE used to solve the circuit.

warning message (Pspice users: Pspice prints "No Convergence During Transient Analysis" in place of this message.) and aborts the simulation. Every other analysis type in SPICE aborts the simulation at the first nonconvergent solution point.

The authors of SPICE recognized that rapid voltage-transition nonconvergence would be a problem during transient analysis. For this reason, the dynamic timestep control algorithm was implemented in SPICE.

Initial conditions

Transient analysis is also different from every other analysis type because of a new initialization technique, the .IC statement. The .IC statement is used to set the initial conditions of the circuit during transient analysis. The syntax of the .IC statement is shown below.

```
.IC V(3)=2.4 V(5)=7.6 V(13)=2.8
```

The .IC statement is similar to the .NODESET statement discussed earlier in this chapter. Both the .IC and the .NODESET are modeled with a Norton-equivalent voltage source. Figure 3.24 illustrates the Norton-equivalent circuit of the .IC statement. But this is where the similarities between the .NODESET statement and the .IC statement end.

The .IC statement is only valid during transient analysis, whereas the .NODESET statement is valid throughout all analysis types. During the DC operating point, DC sweep, and AC frequency sweep analyses, SPICE disregards any .IC statements in the input file. If, however, during a transient analysis, both a .NODESET statement and an .IC statement are included in the input file, only the .IC statement will be used; the .NODESET statement will be discarded.

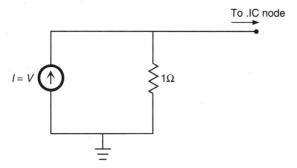

Figure 3.24   The Norton-equivalent voltage source used to model .IC statement voltages in SPICE.

Another difference between the .NODESET and .IC statements involves how SPICE determines the bias point. During a DC analysis, the .NODESET Norton-equivalent power supplies are held in the circuit until a bias point is found. SPICE then removes the power supplies and performs a second DC bias calculation so the Norton-equivalent supplies do not load the actual circuit. Unlike the .NODESET statement, the .IC Norton-equivalent power supplies are held in for only one series of iterations. When a stable bias point has been found, SPICE uses the bias point as the initial transient condition of the circuit. At times, this may cause problems for some circuits because the bias point is established with the Norton-equivalent source still in the circuit.

The .IC statement is often required for setting the initial conditions for *charged* circuit nodes or for starting unstable circuits. Figure 3.25 is a simple RC circuit with a 1V voltage initial condition across the capacitor. To simulate the RC decay of the output voltage, a transient analysis might be performed. Simulate the disk file ch3-26a.cir with the command:

```
SIM CH3-26A.CIR
```

Your results should match Fig. 3.26a. In the ch3-26a.cir circuit file, a .NODESET statement is used to set the initial 1V across the capacitor. The problem with the .NODESET statement is the Norton-equivalent voltage source is removed after a stable bias point has been found. In the ch3-26a.cir, SPICE quickly establishes 1V as the initial DC condition. But after a stable bias point is found, SPICE removes the .NODE-SET voltage source and performs a second series of iterations. In the ch3-26a.cir circuit file, without the Norton-equivalent voltage source, the voltage across the resistor immediately falls to zero. (Remember, the capacitor is not reintroduced into the circuit until after the bias point has been determined.) Because of the need to initialize *charged*

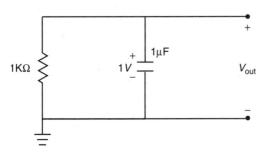

Figure 3.25 A simple circuit to demonstrate the difference between the .NODESET and .IC initial condition statements.

circuit nodes, the authors of SPICE introduced the .IC statement. Simulate the ch3-26b.cir circuit file with the command:

```
SIM CH3-26B.CIR
```

Your results should match Fig. 3.26*b*.

The circuit file ch3-26b.cir is identical to ch3-26a.cir except with an .IC statement in place of the .NODESET statement in the original file.

Nonconvergence in transient analysis

With the differences between the DC sweep analysis and the transient analysis defined, the nonconvergence problems of transient analysis can be examined. The two dominant nonconvergence problems associated with transient analysis are rapid voltage transitions and device model discontinuities. These are the same two failure mechanisms of the DC sweep analysis.

During a DC sweep analysis, rapid voltage-transition nonconvergence was corrected by increasing the number of iterations allowed at each step in the analysis or by reducing the step size of the analysis. In transient analysis, the dynamic timestep control algorithm automatically reduces the step size during large circuit transitions automatically. In a DC sweep analysis, device model discontinuity

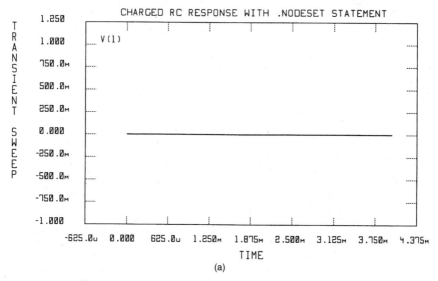

(a)

Figure 3.26(*a*)   The output response of Fig. 3.25 circuit with a .NODESET statement used to set the initial capacitor voltage. (*Reprinted from* Successfully Simulating Circuits with SPICE. *Used with permission.*)

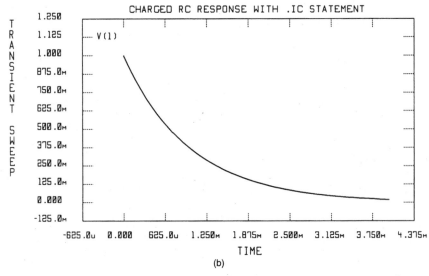

Figure 3.26(b)   The output response of Fig. 3.25 circuit with a .IC statement used to set the initial voltage. (*Reprinted from* Successfully Simulating Circuits with SPICE. *Used with permission.*)

nonconvergence was overcome by increasing the step size but, during transient analysis, the dynamic timestep control algorithm determines the step size.

During transient analysis, as the circuit approaches a voltage transition, two potentially conflicting events occur. First, the timestep control algorithm reduces the size of the timestep, and second, during voltage transitions, the semiconductor devices of the circuit change state and transit from one region of operation into another. As the semiconductor devices change between regions of operation, one or more model discontinuities may be exposed. A voltage transition during transient analysis causes the timestep to be reduced during the same period that device model discontinuities are most likely to be uncovered. Worse yet, if a discontinuity causes a nonconvergence, the timestep control algorithm automatically reduces the size of the timestep, and the reduced timestep is almost guaranteed to stumble into the same discontinuity and fail to converge again. (In the section on DC sweep nonconvergence, model discontinuities were best overcome by increasing the step size to help step over the discontinuity.)

Device model capacitance

Transient nonconvergence is primarily caused by a combination of model discontinuities and a significantly reduced step size brought on by voltage transitions within the circuit. This failure mechanism can be corrected by first recognizing that the model discontinuities are found in the DC I-V characteristics of the device models. All of the device models have a built-in capacitance. The built-in capacitances are either junction capacitance associated with the PN junction of the device or overlap capacitances associated with the insulated gate of a MOSFET. Often these capacitive device currents may help bridge the DC discontinuity. Unfortunately, the built-in capacitances all have a default value of zero, which could lead to convergence problems if a DC discontinuity exists. All real semiconductor devices have some capacitive component; therefore, *all simulation models should have their associated capacitance terms set to a nonzero value.* Adding the capacitance terms to a model helps to improve both the accuracy of the transient analysis and the convergence characteristics of the simulator. Table 3.7 illustrates the capacitance model parameters for each of the semiconductor devices.

If a model parameter set does not include the capacitive terms, either find an accurate value for the capacitance and add the value to the model set, or set the capacitance parameter to a value small enough to be inconsequential to the normal operation of the circuit. Even a small capacitive component is often enough to overcome nonconvergence problems. Table 3.8 illustrates *general* minimum values for each capacitance term for both discrete and integrated circuit components. For some circuits these will be too large. Use Table 3.7 with caution.

A simple SR flip-flop is shown in Fig. 3.27. The flip-flop is constructed from CMOS equivalent gates. The circuit file also contains two pulse voltage sources, one on the set pin and one on the reset pin. The simulation file performs a transient simulation on the flip-flop, driving the latch through several state changes. Simulate the ch3-27.cir file with the command:

```
SIM CH3-27.CIR
```

The circuit will fail to converge at the first voltage transition.

Now simulate the ch3-28a.cir circuit file. This file is identical to the ch3-27.cir file except for the addition of the MOSFET capacitance

TABLE 3.7   Capacitance Model Parameters

Device type	Capacitive model parameters
Diode	CJO
Bipolar transistor	CJE, CJC, CJS
JFET transistor	CGD, CGS
MOSFET transistor*	CGDO, CGSO, CGDO, CBD, CBS, CJ, CJSW

* For MOSFET models, use either CBD and CBS or CJ and CJSW.

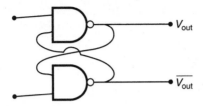

Figure 3.27   A nand-gate flip-flop.

terms CGBO, CGSO, CGDO, CBD, and CBS. The capacitance terms were determined from Table 3.7.

```
SIM CH3-28A.CIR
```

Your result should match Fig. 3.28*a*. In this circuit, adding a small capacitance term eliminates the nonconvergence failure.

Raising the iteration limit

In the section on DC sweep analysis, raising the iteration limit allowed the simulator more iterations before aborting the simulation. For transient analysis, the iteration limit is ITL4. If a timepoint fails to converge within ITL4 iterations, SPICE discards the timepoint, cuts the timestep by a factor of eight, then reattempts the solution with the

TABLE 3.8   Minimum Capacitance Parameter Settings

Device/capacitance parameter	ICs	Discrete
Diode/CJO	.05pF	.1pF
Bipolar/CJE	.1pF	.2pF
Bipolar/CJC	.1pF	.2pf
Bipolar/CJS	.1pF	.2pf
JFET/CGD	.2pF	.2pf
JFET/CGS	.2pF	.2pf
MOSFET/CGDO[†]	20pF	.5pf
MOSFET/CGSO[†]	20pF	.5pf
MOSFET/CGBO[†]	10pF	.025pf
MOSFET/CBD[*]	.1pF	.5pf
MOSFET/CBS[*]	.1pF	.5pf
MOSFET/CJ[*]	20pF	N/A
MOSFET/CJSW[*]	10pF	N/A

[*] For MOSFET models, use either CBD and CBS or CJ and CJSW.
[†] The MOSFET capacitances CGDO, CGSO, and CGBO are per meter of gate length or per meter of gate width. That is why these values appear so much larger than the other capacitive terms.

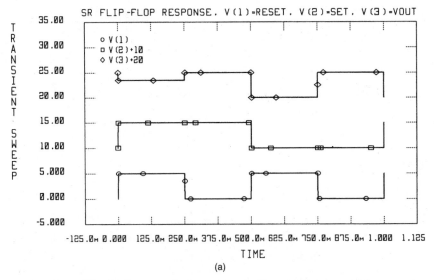

Figure 3.28(a) SR flip-flop transient response with capacitance model parameters included in circuit file. (*Reprinted from* Successfully Simulating Circuits with SPICE. *Used with permission.*)

smaller timestep. By raising ITL4, SPICE has more iterations to converge on a solution. This reduces the number of times SPICE is forced to shrink the timestep; the larger the timestep, the less likely SPICE is to run into or close to the model discontinuity and fail to converge.

By default, ITL4 is set to only 10 iterations. For all transient circuits, raise ITL4 to 40 iterations on the .OPTIONS statement (.OPTIONS ITL4=40).

Simulate the ch3-28b.cir circuit file with the command:

```
SIM CH3-28B.CIR
```

Your results should match Fig. 3.28*b*. The ch3-28b.cir circuit file contains the original SR flip-flop *without the capacitance terms* but with the ITL4 parameter raised to 40. Notice the result obtained with this circuit is identical to the previous result *but the run time for the ch3-28b.cir is less than half of the ch3-28a.cir simulation!*

A side benefit to raising ITL4 is an increase in simulation speed. Chapter 5 will show that the timestep control algorithm in SPICE always cuts the timestep by a factor of 8, yet the timestep control algorithm will increase the timestep by no more than a factor of 2. If SPICE fails to converge and cuts the timestep, at least three more timepoints must be solved before the simulator returns to the point in the simulation where the timestep was originally cut. For many circuits, raising

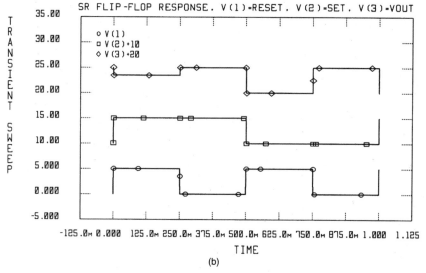

Figure 3.28(b) SR flip-flop transient response with transient iteration limit (ITL4) set to 40. (*Reprinted from* Successfully Simulating Circuits with SPICE. *Used with permission.*)

ITL4 will increase the speed of the simulator and reduce the occurrence of nonconvergence.

Transient convergence aids summary

Overcoming transient nonconvergence failures requires an understanding of DC bias point and DC sweep analysis nonconvergence failures. The dynamic timestep control algorithm helps reduce nonconvergence failures due to rapid voltage transitions but may aggravate failure due to model discontinuities. Restoring the device model capacitance terms from their default of zero to a more realistic value helps minimize model discontinuities. Raising ITL4 gives the timestep algorithm more time to converge on a solution before the timestep is reduced. Maintaining a larger timestep decreases the chance of stepping into or close to a model discontinuity. Used together, these two techniques will eliminate most transient nonconvergence failures. Table 3.9 outlines the transient convergence aids and the order in which they should be applied.

Nonconvergence Summary

Most of the common nonconvergence problems users face can easily be overcome with the controls and options available in SPICE. Many problems occur simply because the error tolerances and program defaults are not appropriate for a given circuit (Table 3.2). Once these

TABLE 3.9   Transient Convergence Aids

Convergence aid	Cause/order
Add capacitance model parameters	First
(.MODEL DIO D (CJO=.1pf)	
Raise ITL4	Second
(.OPTIONS ITL4=40)	

are set, specific analysis types may cause problems. Table 3.5 illustrates the convergence aids for computing the DC bias point of the circuit. Table 3.6 shows the convergence aids for a DC sweep analysis. Finally, the transient convergence aids are shown in Table 3.9.

Set the general convergence aids first. When nonconvergence occurs, determine which analysis caused the nonconvergence, then go to the table for that analysis type. Do not confuse convergence aids. Setting the DC bias point aids will not help resolve transient nonconvergence problems. Setting transient aids will not help during DC sweep. Setting a DC sweep aid will not help during an AC analysis. Identify which analysis caused the nonconvergence, and learn which convergence aids apply to each analysis type.

Convergence Aids
in the SPICE-Like Simulators

Hspice

Hspice has many of the same convergence aids as standard SPICE, as well as several enhancements. Convergence aid options include GMIN, RELTOL, ABSTOL, VNTOL, ITL1, ITL2, and ITL4. The enhancements include Source-Stepping, GMIN-Ramping, and Pseudo-Transient bias determination.

Hspice uses a modified Source-Stepping algorithm when the option CONVERGE=3 is set. Hspice uses a variable step size algorithm in place of the binary stepping algorithm found in SPICE.

When the Hspice option GRAMP=X is specified, Hspice calls the GMIN-Ramping algorithm. GMIN-Ramping involves setting the GMIN conductance (resistance) to a large (small) value. With a large conductance (small resistance) across every nonlinear device in the circuit, much of the nonlinear behavior of the circuit is suppressed. Under these conditions, HSPICE converges very quickly. After converging, Hspice drops the value of GMIN by a factor of 10 and reattempts a solution. This process continues until the circuit fails to converge or GMIN is reduced to the default setting. If Hspice fails to converge, GMIN is increased by a factor of 10 and held at that value for the rest of the simulation.

The GRAMP=X parameter sets the exponential value of the GMIN conductance to start the ramping method. For example, if the .OPTION GRAMP=6 was used, Hspice would ramp the GMIN conductance (resistance) from 1e-6 down to 1e-12 mhos (1e6 ohms up to 1e12 ohms).

To find the DC bias point of the circuit, Hspice added a technique known as Pseudo-Transient analysis. With the Pseudo-Transient analysis, a small capacitor is attached to every node in the circuit. The analysis continues by ramping the DC power supplies from zero up to full power. Once the power supplies are at full power, the capacitors are removed from the circuit, and the result is the proper DC bias point. The Hspice option CONVERGE=1 invokes the Pseudo-Transient analysis.

IS_Spice

IS_Spice has the same convergence aid as standard SPICE. Convergence aid options include GMIN, RELTOL, ABSTOL, VNTOL, ITL1, ITL2, ITL3, ITL4, and ITL6.

Micro-Cap IV

Micro-Cap IV has many of the same convergence aids as standard SPICE, with the addition of the SAVE BIAS command. Convergence aid options include GMIN, RELTOL, ABSTOL, VNTOL, ITL1, ITL2, ITL3, and ITL4.

While not directly a convergence aid, the SAVE BIAS command allows users to reuse the bias point voltages from a previous simulation. For large circuits, reusing the bias point may increase the speed of subsequent runs by eliminating the time and iterations consumed in the bias point calculation.

Pspice

Pspice has many of the same convergence aids as standard SPICE, with three additional enhancements. Convergence aid options include GMIN, RELTOL, ABSTOL, VNTOL, ITL1, ITL2, and ITL4. The convergence enhancements include an additional GMIN resistor across the independent current sources and dependent voltage and current sources, automatic source-stepping for the DC bias point and DC sweep calculations, a reduced resistor for the .NODESET and .IC statements, and the SAVEBIAS command.

A GMIN resistor is placed across each independent current source and every dependent source in the circuit. Changing the value of gmin on the .OPTIONS statement alters this value.

If Pspice fails to converge within ITL1 iterations during a DC bias point calculation, or ITL2 iterations during a DC sweep analysis, the program automatically switches into the Source-Stepping algorithm. The Source-Stepping algorithm in Pspice uses a variable step size rather than the binary stepping method of standard SPICE.

In SPICE, .NODESET and .IC statements are modeled with Norton-equivalent source. The Norton-equivalent source contains a current source and a 1-ohm resistor. For many low-impedance circuits, the 1-ohm resistor adds significant loading to the circuit. In Pspice, the 1-ohm resistor is replaced with a .002-ohm resistor, and the value of the current source is increased to match the reduced resistance.

While not directly a convergence aid, the SAVEBIAS command allows users to reuse the bias point voltages from a previous simulation. For large circuits, reusing the bias point may increase the speed of subsequent runs by eliminating the time and iterations consumed in the bias point calculation.

Summary

Nonconvergence is one of the most common and frustrating problems facing simulation users. Resolving nonconvergence problems involves identifying and eliminating the cause of the nonconvergence. Set the general convergence aids on every circuit you simulate. If nonconvergence occurs, identify which analysis type caused the nonconvergence and set the analysis-specific convergence aids. Following the guidelines outlined here will resolve between 80 and 90 percent of all nonconvergence problems.

Some nonconvergence problems may be more stubborn than others. For tough problems, follow the guidelines exactly as shown. Sometimes, the first convergence aid you try will resolve the problem. Other problems will require all of the convergence aids. Do not worry about adding too many convergence aids. You will not introduce any appreciable error in your simulation *if you follow the guidelines offered in this chapter.* The guidelines presented here represent a proven, systematic approach to resolving nonconvergence problems.

Overcoming convergence problems is possible and practical. Before the accuracy of a simulation can be determined, before the speed of a simulation can be improved, before the analysis results can be examined, *the circuit must converge.*

References

1. RCG Research's Successfully Simulating Circuits with SPICE Training Class, SPICE Intermediate, Nonconvergence Section, 1990.
2. J. Vlach, K. Singhal, *Computer Methods for Circuit Analysis and Design,* Van Nostrand Reinhold, 1983, ISBN 0-442-28108-0.

Numeric Integration

What is numeric integration? Why is numeric integration used in SPICE? These are two important questions which any knowledgeable simulation user should be able to answer. If you don't know the answers to these questions, read on. Too few users realize the significant role that numeric integration plays in producing an accurate transient simulation.

Numeric integration is used in SPICE to calculate the current flowing through the circuit capacitors as a function of time; in a dual role, numeric integration is used to calculate the voltage across the circuit inductors as a function of time.

As to what numeric integration is, numeric integration is analytic integration applied in a piecewise linear manner. Whereas analytic integration produces a continuous function which represents the solution, numeric integration produces discrete solution points. If all the solution points are plotted on a graph, the resulting curve will match the function generated from the analytic integration.

Most often, numeric integration is used to solve differential equations which have an unknown or difficult-to-calculate analytic solution. In the case of circuit simulation, the differential equations of interest are:

$$I = \frac{C*dV}{dt} \qquad (4.1)$$

and

$$V = \frac{L*dI}{dt} \qquad (4.2)$$

Numeric integration determines the capacitor currents and inductor voltages for a transient simulation by solving the differential equations relating voltage and current.

During transient analysis, SPICE uses one of three user-selectable integration methods—trapezoidal, backward-Euler, or Gear—to determine the capacitor currents and inductor voltages. In addition to the standard Gear integration formula, four additional variations of the Gear equation may be selected. This gives the SPICE user a wide variety of integration methods from which to choose.

Transient Simulation Warning

A variety of integration methods are provided in SPICE simply because *no one integration method is always the best for all transient simulations*. When a numeric integration formula is applied to a differential equation, the piecewise linear solution generated by the integration is only an approximation of the exact function. At each solution point, the integration algorithm may introduce a small amount of error. The accuracy of the numeric integration method will depend on the size of the simulation timestep, the shape of the voltage and current waveforms, and the integration method being used. Because of the many different waveforms which arise during transient analysis, the numeric integration algorithms must be extremely robust to maintain accuracy throughout the variety of circuit responses. A given integration method may work well on one function or waveform and produce completely inaccurate results on another. Because no one integration method is best suited for all transient simulations, SPICE users must be able to decide which integration method produces the most accurate result for their circuit.

When used without caution, numeric integration inaccuracies can be an Achilles' heel for the simulation user. Unlike nonconvergence failures, *SPICE will not generate a warning message when the numeric integration method is producing inaccurate results.* The SPICE user must learn to look for signs which indicate something has gone wrong during the simulation. Often, when the integration method does fail, simulation results will display well-defined, well-characterized anomalous behavior. A numeric integration failure may appear as pronounced as a catastrophic failure or as subtle as a voltage or current ripple. But to the trained eye, these same signs are the signals to switch to a different integration method. Learning the characteristic signs of numeric integration failure, and what to do to correct the failure, are important lessons in simulation craftsmanship.

Warning summary

Simulation users should not presume the results obtained from their simulations are always correct. There are two reasons for this: first, model inaccuracies will lead the simulator to inaccurate results; second, even when the models are completely accurate, several failure mechanisms of the simulator can lead to the wrong solution (numeric integration failure is just one of the failure mechanisms) without warning the user. *Never simply assume the simulation output is correct.* If questionable results appear, use good engineering judgment to determine whether the anomaly was simulator-related or design-related. Never simulate a circuit without a reasonably good idea of what the simulation output should be. A good designer would never breadboard a collection of resistors, capacitors, inductors, and transistors without some reasonable expectation of the behavior of the circuit. Simulation users should treat computer simulation the same way.

Numeric Integration

Before going further, a simple example of numeric integration might be useful here. A time function is shown in Eq. 4.3.

$$F(t) = 2t \qquad (4.3)$$

The derivative of Eq. 4.3 may easily be computed and is shown in Eq. 4.4. The derivative $F'(t)$ is a constant and is by definition a differential equation. Equation 4.4 will serve as the starting point for our example.

$$F'(t) = 2 \qquad (4.4)$$

Equation 4.4 is a differential equation. A differential equation is nothing more than the derivative of a function. Often the original function is unknown, but in this case both the derivative and the original function are known. Solving a differential equation involves transforming the differential equation back into the original function.

Equation 4.4 may be solved with an analytic technique known as Separation of Variables. Begin by rewriting Eq. 4.4 in the form of a differential equation.

$$\frac{dY(t)}{dt} = 2 \qquad (4.5)$$

Multiplying both sides of Eq. 4.5 by dt results in Eq. 4.6.

$$dY(t) = 2*dt \qquad (4.6)$$

Equation 4.6 may be integrated over time, yielding Eq. 4.7.

$$Y(t) = 2t + C \qquad (4.7)$$

Not surprisingly, Eqs. 4.7 and 4.3 are identical except for the constant C.

The technique of separation of variables is an analytic method of solving a differential equation. Equation 4.5 can also be solved with numeric integration.

Numeric integration produces results *similar* to the analytic solution but in a piecewise linear form. Notice the word "similar" was used, not "identical." Numeric integration is an approximation to the solution. At each solution point, a finite amount of error may be introduced. Often, the error will be insignificant and the numeric integration solution will be very close to the analytic solution, but at times the integration error can be significant.

During numeric integration, the original function is reconstructed as a series of discrete data points at each step in the integration. To apply numeric integration to Eq. 4.5, construct an X-Y coordinate grid as shown in Fig. 4.1. Since Eq. 4.4 is a function of *time,* discrete units of time will be the interval of integration.

To apply numeric integration to a differential equation, an integration formula must be chosen, and a starting point or initial value of the solution must be known. For this example, the backward-Euler numeric integration formula will be used, and $Y(t = 0)$ will be set to zero. The backward-Euler formula is shown in Eq. 4.8.

$$Y(x + 1) = Y(x) + \text{step size} * \frac{dY(x + 1)}{dx} \qquad (4.8)$$

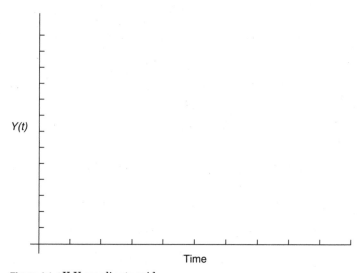

Figure 4.1 X-Y coordinate grid.

Equation 4.9 results when the backward-Euler formula is applied to Eq. 4.5.

$$Y(t + 1) = Y(t) + \text{timestep} * \frac{dY(t + 1)}{dt} \qquad (4.9)$$

In Eq. 4.9, $Y(t + 1)$ represents the solution to the differential equation at the timepoint $T + 1$. $Y(t)$ is the solution at the past timepoint T. For our example, the timestep will be set to a constant .5 seconds. The integration may be started by setting $Y(t = 0)$ to zero and evaluating Eq. 4.9 at discrete .5-second timepoints. Table 4.1 displays the calculation and the computed value of $Y(t)$ at each timepoint. Figure 4.2 reconstructs the graph of $Y(t)$.

Many practical engineering problems must be solved with differential equations. The goal of numeric integration is to reconstruct the original function from the known differential equation. At each step in the integration, the integration formula produces a discrete solution value. The original function is represented by the set of solution values.

In this example, the numeric integration solution was exactly equal to the analytic solution, but only because this was a simple function where the differential equation is constant. In more complex functions, such as those with a nonconstant first or nonzero second derivative, many of the numeric integration methods will produce small amounts of error at each step in the integration.

SPICE and Numeric Integration

In SPICE, the numeric integration routines solve the differential equations describing the current-voltage relationships for capacitors and inductors. Equation 4.10a is the backward-Euler expression for the I-V

TABLE 4.1 Calculated Timepoint Solution Values

Y(0)	= 0		
Y(.5)	= 0	+ .5 * 2	= 1
Y(1.0)	= 1	+ .5 * 2	= 2
Y(1.5)	= 2	+ .5 * 2	= 3
Y(2.0)	= 3	+ .5 * 2	= 4
Y(2.5)	= 4	+ .5 * 2	= 5
Y(3.0)	= 5	+ .5 * 2	= 6
Y(3.5)	= 6	+ .5 * 2	= 7
Y(4.0)	= 7	+ .5 * 2	= 8
Y(4.5)	= 8	+ .5 * 2	= 9
Y(5.0)	= 9	+ .5 * 2	= 10

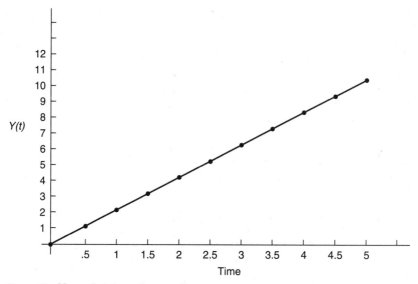

Figure 4.2   Numeric integration result.

relationship of a capacitor. A dual expression can be written for the inductor.

$$V_c(t + 1) = V_c(t) + \frac{\text{timestep} * I_c(t + 1)}{C} \qquad (4.10a)$$

To determine the capacitor current from Eq. 4.10a, the voltage and current terms must appear on opposite sides of the equality. Using simple algebra to rearrange the terms of Eq. 4.10a results in a more familiar expression for capacitor current.

$$\frac{[V_c(t + 1) - V_c(t)]}{\text{timestep}} * C = I_c(t + 1) \qquad (4.10b)$$

At each timepoint in the transient simulation, the node voltages from the present solution point $V_c(t + 1)$ and the previous timepoint $V_c(t)$ are used to determine the change in capacitor voltage.

$$\frac{[V_c(t) - V_c(t - 1)]}{\text{timestep}} = \frac{dV_c}{dt} \qquad (4.11)$$

Once the change in voltage is known, the capacitor current can be determined by multiplying Eq. 4.11 by the capacitance value.

$$\frac{C * [V_c(t + 1) - V_c(t)]}{\text{timestep}} = \frac{C * dV_c}{dt} = I_c(t + 1) \qquad (4.12)$$

When SPICE begins the transient simulation, an equation similar to Eq. 4.12 is generated for every capacitor in the circuit, and a dual equation is generated for every inductor in the circuit.

In searching for a solution, SPICE iterates on node voltage values over and over again until a set of voltages is found that satisfies Kirchhoff's voltage and current laws and satisfies Eq. 4.12 for every capacitor/inductor in the circuit. Since $V_c(t)$ is the node voltage from the previous timepoint (not from the previous iteration) and remains fixed, only $V_c(t + 1)$ changes with each new iterative voltage value. As $V_c(t + 1)$ changes, $I_c(t + 1)$ changes to match the new capacitor voltage. When the exact solution is found, Eq. 4.12 becomes the relationship between the capacitor voltage and current at the transient timepoint.

Types of Numeric Integration

SPICE offers users three different numeric integration methods for use during transient simulation. The three methods are backward-Euler, trapezoidal, and Gear integration. Of these, both the backward-Euler and trapezoidal methods were developed in the mid-1700s.[1] Only the Gear method was developed during this century.[2] The different integration methods were written into SPICE because Nagel[3] found that no one method was reliable under all transient simulations. For some circuits the trapezoidal method produces the most accurate result; for other circuits, the Gear method works best; and yet other circuits will produce the best result with the backward-Euler method.

Although each integration method must perform the same task, the way each method generates the solution will differ. An understanding of how each method accomplishes the task of solving the capacitor currents and inductor voltages, and knowing the difference between methods, will help in choosing an appropriate method for a given circuit type.

Although SPICE has only three different integration types, four will be presented here to illustrate the differences and similarities between methods. The four of interest are forward-Euler, backward-Euler, trapezoidal, and Gear integration. To illustrate the differences between methods, a graphical technique utilizing two X-Y coordinate grids will be employed. The Y-axis of the first graph represents the differential equation $dY(t)/dt$. The second graph represents both the exact analytic solution and the numeric integration solution. The analytic solution is shown in the second graph as a solid line. The numeric integration solution is shown in the graph as a series of data points.

Figure 4.3 represents the forward-Euler integration method. The forward-Euler estimates the value of the function $Y(t + 1)$ by adding the previous solution value $Y(t)$ to the product of the timestep and the derivative of the timepoint T, $dY(t)/dt$. The name forward-Euler comes from the use of the previous timepoint T in establishing the derivative for a solution at $T + 1$. In a sense, the forward-Euler method is estimating the solution based on the derivative of the past solution point. The forward-Euler integration formula is shown in Eq. 4.13.

$$Y(t + 1) = Y(t) + \text{timestep}*dY(t) \qquad (4.13)$$

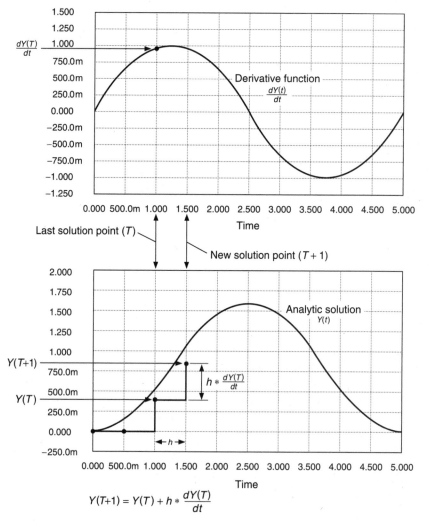

$$Y(T+1) = Y(T) + h * \frac{dY(T)}{dt}$$

Figure 4.3 Forward-Euler numeric integration formula and graphical interpretation of method.

Figure 4.4 illustrates the backward-Euler integration method. The forward- and backward-Euler methods are very similar. The difference between the two methods is the selection of the timepoint where the derivative of the function is being evaluated. The backward-Euler method estimates the value of the function $Y(t + 1)$ by adding the previous solution value $Y(t)$ to the product of the timestep and the derivative of the timepoint $T + 1$, $dY(t + 1)/dt$. Whereas the forward-Euler method uses the derivative at the past solution point T, the backward-

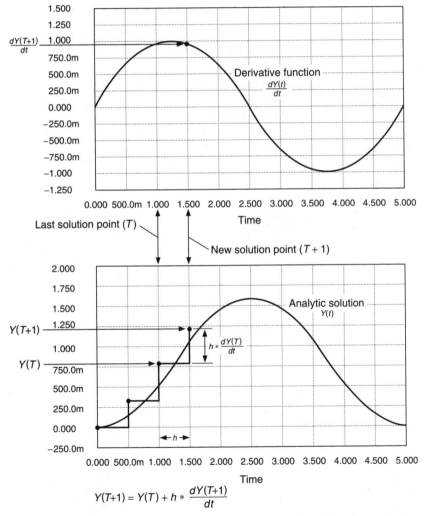

$$Y(T+1) = Y(T) + h * \frac{dY(T+1)}{dt}$$

Figure 4.4   Backward-Euler numeric integration formula and graphical interpretation of method.

Euler method uses the derivative of the current solution point $T + 1$. The backward-Euler integration formula is shown in Eq. 4.14.

$$Y(t + 1) = Y(t) + \text{timestep} * dY(t + 1) \qquad (4.14)$$

Figure 4.5 represents the trapezoidal integration method. The trapezoidal is similar to the Euler methods but again differs in the selection of where the derivative of the function will be evaluated. Like the Euler methods, the trapezoidal method estimates the value of $Y(t + 1)$ by adding the previous solution value $Y(t)$ to the product of the timestep and an *averaged* derivative. The trapezoidal algorithm uses the derivative of the past solution point T and the derivative at the current solution point $T + 1$. These two derivative values are averaged, and this average derivative is used in the calculation. Rather than evaluating the derivative at the past timepoint (forward-Euler) or at the present timepoint (backward-Euler), the trapezoidal method averages the two derivatives as shown in Fig. 4.5. Because of this, the trapezoidal method tends to be more accurate than either the forward- or backward-Euler methods. The trapezoidal integration formula is shown in Eq. 4.15.

$$Y(t + 1) = Y(t) + \frac{\text{timestep} * [dY(t) + dY(t + 1)]}{2} \qquad (4.15)$$

The Gear integration method[2] is a relative newcomer. Figure 4.6 represents the Gear integration method. The Gear integration method is very different from traditional integration methods and estimates the value of the function at time $T + 1$ with information from the present solution point and two past solution points! The Gear-2 integration formula is shown in Eq. 4.16.

$$Y(t + 1) = \left(\frac{4}{3}\right) * Y(t)$$

$$- \left(\frac{1}{3}\right) * Y(t - 1)$$

$$+ \left(\frac{2}{3}\right) * \text{timestep} * dY(t + 1) \qquad (4.16)$$

In addition to the Gear-2 formula, there are Gear-3, Gear-4, Gear-5, and Gear-6 formulas. Where the Gear-2 formula uses two previous timepoints, $Y(t)$ and $Y(t - 1)$, the Gear-3 formula uses three, $Y(t)$, $Y(t - 1)$, and $Y(t - 2)$. This progression continues up to the Gear-6 formula which uses six previous timepoints to help estimate where the function will be at time $T + 1$.

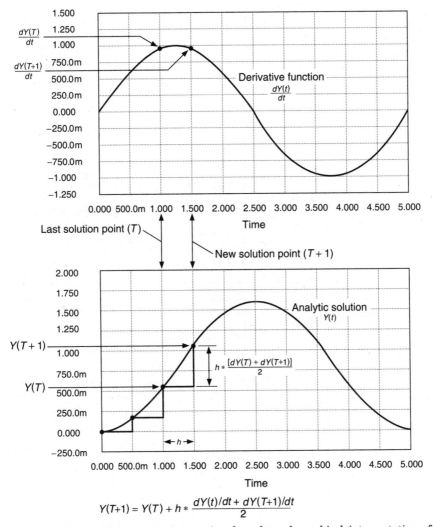

$$Y(T{+}1) = Y(T) + h * \frac{dY(t)/dt + dY(T{+}1)/dt}{2}$$

Figure 4.5 Trapezoidal numeric integration formula and graphical interpretation of method.

Accuracy and Stability of Integration Methods

Due to the difference in the integration formulas, each method will produce a different result when applied to a given function. How well a given integration formula will predict the correct solution is determined by the accuracy and stability of the method.

An integration method is approximating the solution of a differential equation at each timepoint in the transient simulation. Because the

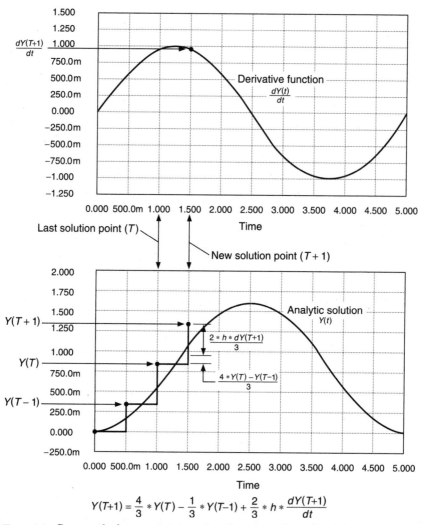

Figure 4.6 Gear method numeric integration formula and graphical interpretation of method.

numeric integration solution is only an approximation to the exact solution value, a finite amount of error may be introduced at each time-point. The amount of error at a given timepoint is a measure of the *accuracy* of the integration method. Of the four integration methods presented (backward-Euler, forward-Euler, trapezoidal, and Gear), each will have a different amount of error introduced at a given time-point. The error introduced at each timepoint in a numeric integration is known as the local truncation error. Figure 4.7 illustrates the accuracy of an integration method and the local truncation error for an arbitrary function.

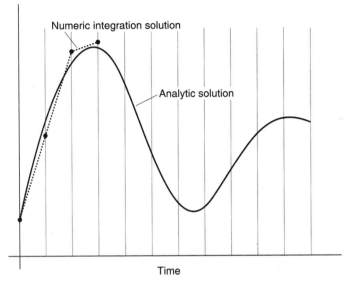

Figure 4.7   The accuracy of a numeric integration method refers to the amount of error at each timepoint of the integration. The error at each timepoint is known as Local Truncation Error.

A second criterion for a numeric integration method is stability. How the local truncation error accumulates over a large number of time-points is a measure of the *stability* of an integration method. A stable integration method may generate a result which overestimates the actual solution at some timepoints and underestimates the actual solution at others. But over a large number of timepoints (a long transient simulation), a stable integration method produces a result which closely approximates the actual solution. Conversely, an unstable method tends to accumulate the local truncation error at each time-point. Over a large number of timepoints, an unstable integration method will diverge from the exact solution. Figure 4.8 shows the result of a stable and an unstable integration method when applied to a general function.

Both the accuracy and stability of a given integration method will be determined by the first- and second-order derivatives of the function being integrated and by the timestep used during integration. Because of this, no one integration method will always be the most accurate or the most stable. An integration method may generate large amounts of local truncation error when applied to one function, and generate little or no error when applied to a different function. A given integration method may be stable on some functions and unstable on others.

Although this text will set some general guidelines on which integration method you should use on your circuits, any integration

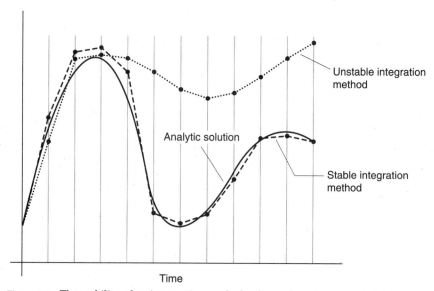

Figure 4.8 The stability of an integration method refers to how the numeric integration solution behaves over a large number of timepoints. Accurate numeric integration methods may not always be stable and stable methods may not be accurate.

method is prone to accuracy and stability problems. A knowledgeable simulation user will learn to recognize numeric integration inaccuracies and will take corrective actions. The following sections will show you how to look for anomalous behavior and what to do to correct your simulation.

Comparison of Integration Methods

To illustrate the difference between integration methods, three sample time functions were chosen. All three had smooth and continuous derivatives. The derivative of each function was calculated and then the forward-Euler, backward-Euler, trapezoidal, and Gear-2 integration methods were applied to the computed derivatives. The resulting curves represent the four numeric integration approximations of the original function.

Equations 4.17, 4.18, and 4.19 show the three sample functions used for this illustration.

$$f(t) = t^2 - 5t \qquad (4.17)$$

$$f(t) = 5 - 5*exp(-t) \qquad (4.18)$$

$$f(t) = \sin(t) \qquad (4.19)$$

Figures 4.9*a, b,* and *c* represent Eq. 4.17 for a step size of 1s, .5s, and .25s respectively. By decreasing the step size of the integration, the local truncation error decreases. While no one integration method is always the most accurate or the most stable, *decreasing the step size of any integration method improves the accuracy of the solution.* Figures 4.10*a, b,* and *c* represent Eq. 4.18, and Figs. 4.11*a, b,* and *c* represent Eq. 4.19. In each illustration, decreasing the step size produces a pronounced improvement in the accuracy of solution.

The previous statement deserves careful consideration. By decreasing the step size of the integration, or by decreasing the timestep during transient analysis, the resulting numeric integration solution for capacitor currents and inductor voltages improves. Many SPICE users know that decreasing the timestep cures many transient anomalies. (A more accurate numeric integration solution is only one of several improvements brought on by a smaller transient timestep.) But decreasing the timestep may accentuate two undesirable characteristics of the simulator. Decreasing the timestep forces the simulator to solve more solution points and results in longer simulation runs, and decreasing the timestep tends to increase the chance of stepping into or close to a model discontinuity and failing to converge. *In any simulator, a delicate balance exists between speed, accuracy, and the ability to converge.* This natural tension may be seen when attempting to select a

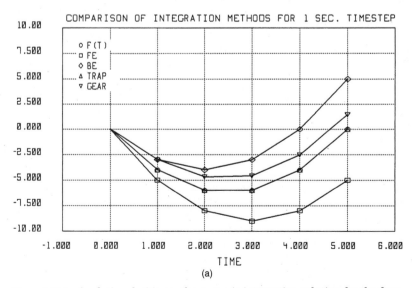

Figure 4.9(a)   Analytic solution vs. the numeric integration solution for the function $F(t) = t^2 - 5t$. (*Reprinted from* Successfully Simulating Circuits With SPICE. *Used with permission.*)

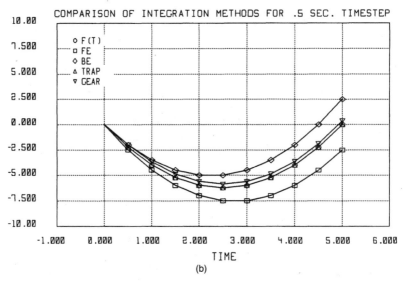

Figure 4.9(b) Analytic solution vs. the numeric integration solution for the function $F(t) = t^2 - 5t$. (*Reprinted from* Successfully Simulating Circuits With SPICE. *Used with permission.*)

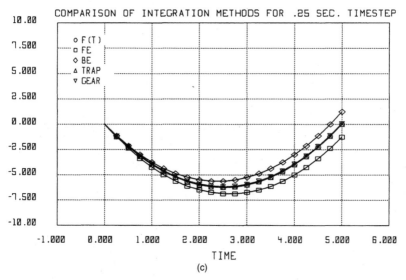

Figure 4.9(c) Analytic solution vs. the numeric integration solution for the function $F(t) = t^2 - 5t$. (*Reprinted from* Successfully Simulating Circuits With SPICE. *Used with permission.*)

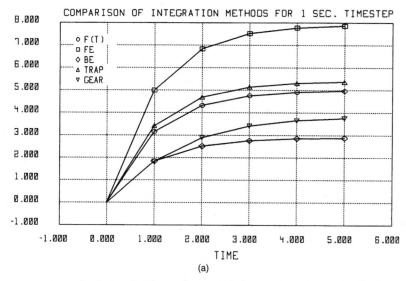

Figure 4.10(a) Analytic solution vs. the numeric integration solution for the function $F(t) = 5 - 5e^{-t}$. (*Reprinted from* Successfully Simulating Circuits With SPICE. *Used with permission.*)

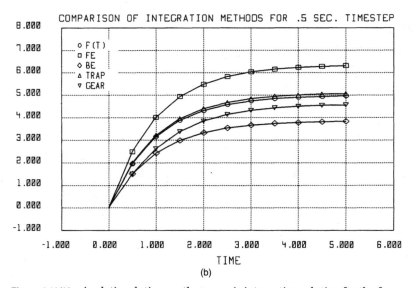

Figure 4.10(b) Analytic solution vs. the numeric integration solution for the function $F(t) = 5 - 5e^{-t}$. (*Reprinted from* Successfully Simulating Circuits With SPICE. *Used with permission.*)

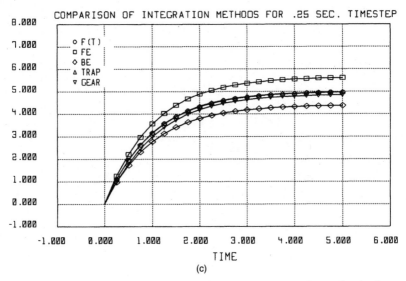

Figure 4.10(*c*)   Analytic solution vs. the numeric integration solution for the function $F(t) = 5 - 5e^{-t}$. (*Reprinted from* Successfully Simulating Circuits With SPICE. *Used with permission.*)

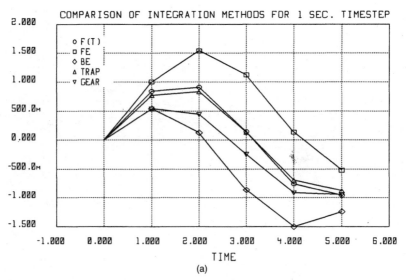

Figure 4.11(*a*)   Analytic solution vs. the numeric integration solution for the function $F(t) = \sin(t)$. (*Reprinted from* Successfully Simulating Circuits With SPICE. *Used with permission.*)

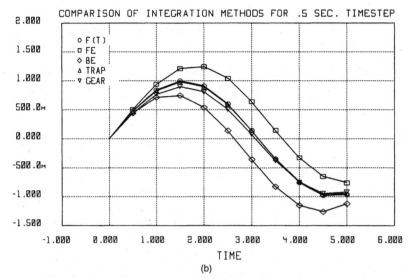

(b)

Figure 4.11(b)   Analytic solution vs. the numeric integration solution for the function $F(t) = \sin(t)$. (*Reprinted from* Successfully Simulating Circuits With SPICE. *Used with permission.*)

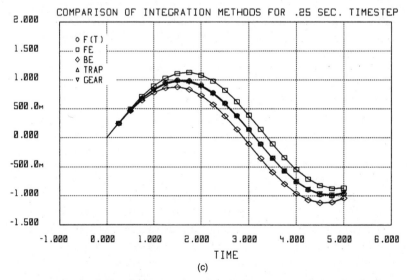

(c)

Figure 4.11(c)   Analytic solution vs. the numeric integration solution for the function $F(t) = \sin(t)$. (*Reprinted from* Successfully Simulating Circuits With SPICE. *Used with permission.*)

timestep for transient analysis which maintains accuracy, reasonable convergence characteristics, and reasonable simulation time.

Numeric Integration of Electronic Circuits

If we narrow our focus to the general class of functions most often seen in electronic circuits (linear, piecewise linear, exponential, and sinusoidal), some *general* observations may be made about numeric integration.

First, the forward-Euler method is generally inaccurate and unstable on these types of functions, and therefore is not used in SPICE. Second, remaining methods—backward-Euler, trapezoidal, and Gear—may be represented on a continuum with accuracy on the left-most side and stability on the right-most side as shown in Fig. 4.12*a*. The trapezoidal is generally the most accurate; the Gear methods, because of their smoothing or averaging function, tend to be the most stable. (The backward-Euler method is sometimes referred to as a Gear-1 integration method although no formal Gear-1 method exists.) But too much stability is not always good. On highly nonlinear circuits, past solution values may yield extremely poor estimates of future solution values. In general, the Gear-3, Gear-4, Gear-5, and Gear-6 integration methods show little or no improvement over the Gear-2 integration method, and, since these methods are more complex and involve more computational steps, resulting in longer simulation runs, the Gear-3, Gear-4, Gear-5, and Gear-6 methods are not recommended for use in simulation. For general electronic circuit simulation, backward-Euler, trapezoidal, and Gear-2 integration methods will produce the best results. These three methods are shown in Fig. 4.12*b*.

Before we go further, it is important to take note of the difference between Eqs. 4.10*a* and 4.10*b*. Although these two equations are essentially the same, *a* is the equation of an integrator, while *b* is the equation of a differentiator. Integration tends to be a smoothing or averaging function, while differentiation tends to be a noisy and sometimes erratic function. Although known as numeric integration, SPICE reverses the equality in the integration formula and uses numeric *differentiation* to determine the capacitor currents and inductor voltages.

The step response is one measure of the stability of a system. By applying a step function to the numeric integration routines, several very important observations may be made. In Fig. 4.13, a ramp voltage is applied to a 1-farad capacitor. The voltage increases at a constant 1 volt per second. At time $t = 1$, the voltage is held constant. Using analytic integration, the capacitor current can be calculated directly from Eq. 4.1. Both the capacitor voltage and current are illustrated in Fig. 4.13.

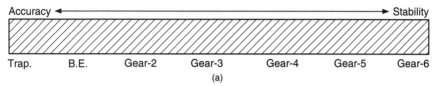

Accuracy ◄──────────────────────────────────────► Stability

Trap.   B.E.   Gear-2   Gear-3   Gear-4   Gear-5   Gear-6

(a)

Figure 4.12(a)   Continuous exchange between accuracy and stability.

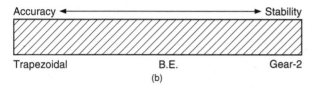

Accuracy ◄──────────────────────────────► Stability

Trapezoidal   B.E.   Gear-2

(b)

Figure 4.12(b)   Suggested integration methods for SPICE simulation.

By replacing the analytic integration with a numerical integration, we can compare how each integration method performs when used as a numeric differentiator.

Figures 4.14*a* and *b* illustrate the capacitor current response generated with the application of the backward-Euler integration method. In Fig. 4.14*a* the backward-Euler integration method was applied with a fixed step size of .025S. Figure 4.14*b* is the same calculation with a fixed step size of .0025s. The smaller step size improves the integration result but takes longer to calculate.

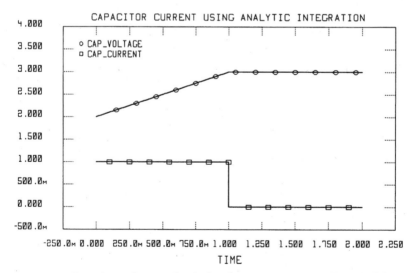

Figure 4.13   Capacitor voltage and calculated current response. (*Reprinted from Successfully Simulating Circuits With SPICE. Used with permission.*)

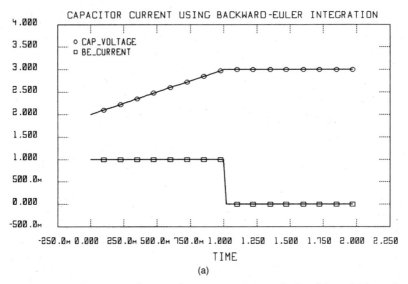

(a)

Figure 4.14(a)   Capacitor voltage and current response calculated from backward-Euler integration and a step size of .025S. (*Reprinted from* Successfully Simulating Circuits With SPICE. *Used with permission.*)

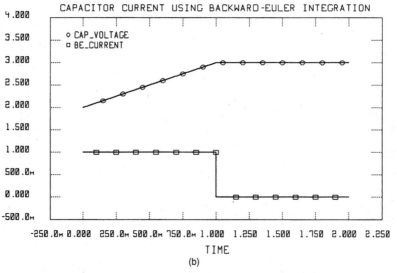

(b)

Figure 4.14(b)   Capacitor voltage and current response calculated from backward-Euler integration and a step size of .0025S. (*Reprinted from* Successfully Simulating Circuits With SPICE. *Used with permission.*)

Figures 4.15*a* and *b* illustrate the capacitor current response generated with the use of the trapezoidal integration method. In Fig. 4.15*a* the capacitor current after time *t* = 1 appears to oscillate around the correct value of zero. While this result may surprise many readers, the effect of *ringing* is a well-known trapezoidal failure mechanism. In many functions which contain an abrupt change or discontinuous derivative, *dV/dt*, using the trapezoidal integration method *with a large timestep* results in the phenomenon known as *trapezoidal oscillation*. The oscillation occurs because of the eigenvalues[5] of the circuit and the size of the timestep being used. The oscillation does not occur often but will occur when the eigenvalues of the circuit are small and the timestep is large. Trapezoidal oscillation is not a result of components in the circuit; the oscillation is simply an erroneous result of the numeric integration algorithm. Trapezoidal oscillation is a readily observable clue which indicates the numeric integration method is failing to predict the correct response, and corrective actions should be taken.

Of the three integration methods available in SPICE, the trapezoidal method is the only one which exhibits this oscillatory behavior. For most functions, the oscillation can be suppressed or eliminated by reducing the step size of the integration. By reducing the step size to .0025S, the oscillation seen in Fig. 4.15*b* is reduced significantly from the previous illustration. Reducing the step size further will eliminate the oscillation.

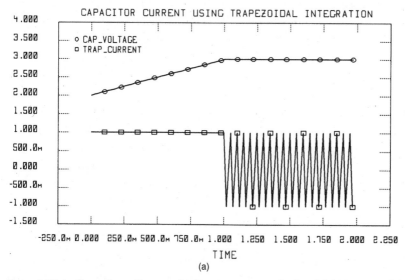

Figure 4.15(a)   Capacitor voltage and current response calculated from trapezoidal integration and a step size of .025S. (*Reprinted from* Successfully Simulating Circuits With SPICE. *Used with permission.*)

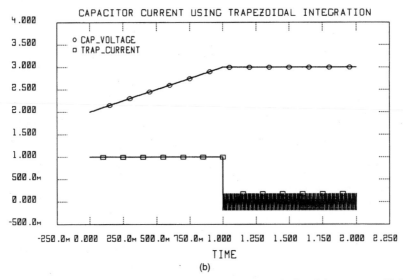

Figure 4.15(b)   Capacitor voltage and current response calculated from trapezoidal integration and a step size of .0025S. (*Reprinted from* Successfully Simulating Circuits With SPICE. *Used with permission.*)

SPICE employs a dynamic timestep control algorithm and source-breakpoint adjustment (these will be discussed in detail in Chap. 5) to help minimize trapezoidal oscillation. For many simulations these techniques work well at controlling the oscillation, but, for some circuits, trapezoidal oscillation will be readily observable.

Figures 4.16a and b illustrate the capacitor current response generated with the application of the Gear-2 integration method. In Fig. 4.16a the capacitor current after time $t = 1$ appears to overshoot the correct value of zero. This overshoot is a well-known characteristic of the Gear integration method. Because the Gear integration method uses past timepoints to help determine the next solution point, circuit functions with an abrupt change often display a result which appears to overshoot the correct response. As in the case of trapezoidal oscillation, the Gear-induced overshoot is not the result of components in the circuit; the overshoot is simply an error introduced by the integration algorithm. The Gear-2 integration method is the most stable integration method in SPICE. However, the same mechanism which adds to the stability of the Gear method tends to resist abrupt changes. This resistance to change produces the overshoot seen in Fig. 4.16a.

One way to reduce or eliminate the overshoot is to again reduce the size of the timestep being used during the integration. Figure 4.16b illustrates how a much smaller step size reduces the overshoot at time $t = 1$.

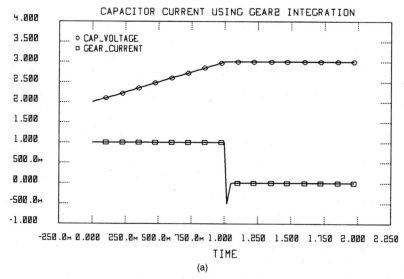

Figure 4.16(a)   Capacitor voltage and current response calculated from Gear-2 integration and a step size of .025S. (*Reprinted from* Successfully Simulating Circuits With SPICE. *Used with permission.*)

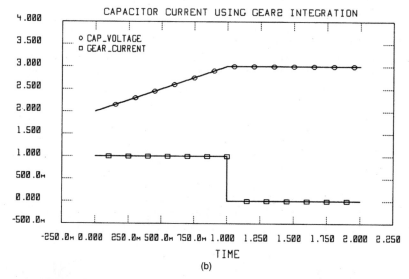

Figure 4.16(b)   Capacitor voltage and current response calculated from Gear-2 integration and a step size of .0025S. (*Reprinted from* Successfully Simulating Circuits With SPICE. *Used with permission.*)

Trapezoidal oscillation and Gear overshoot seen in the previous illustrations are two basic failure mechanisms which may contribute inaccuracies to your transient simulation. These same characteristics are the kind of clues which users need to watch for during transient simulation. SPICE will not flag these as anomalous behaviors. As a SPICE user, you must observe these behaviors and make the appropriate adjustments to your simulation.

Selecting an integration method

Because SPICE uses only one integration method for the entire circuit, simulation users need to choose the best method for their circuit. The authors of SPICE originally chose the trapezoidal integration as the default integration method. The trapezoidal method is a good choice as the default method because, on the type of functions most often seen in electronic circuits, the trapezoidal tends to produce the most accurate result. In addition to the algorithm's accuracy, the trapezoidal method is also faster to evaluate than the Gear integration methods. For many types of circuits, the trapezoidal integration method is the best choice for simulation.

But the trapezoidal integration method is not *always* the best choice. In addition to the trapezoidal integration method, the authors of SPICE also implemented the backward-Euler and the Gear integration methods and gave simulation users the option of selecting which integration method to use.

Changing integration methods

SPICE users may select the trapezoidal, backward-Euler, or Gear methods of integration. The switch to change integration methods is appropriately found in the .OPTIONS statement. To select either Gear-2 or trapezoidal integration methods, use the following option:

```
.OPTIONS METHOD=TRAP    (default)
.OPTIONS METHOD=GEAR
```

If the METHOD=GEAR option is chosen, SPICE will use the Gear-2 integration method. Although this text does not recommend using the higher-order Gear methods (orders higher than 2), the higher-order Gear methods may be selected with the MAXORD option as shown below:

```
.OPTIONS METHOD=GEAR MAXORD=3
.OPTIONS METHOD=GEAR MAXORD=4
.OPTIONS METHOD=GEAR MAXORD=5
.OPTIONS METHOD=GEAR MAXORD=6
```

To select the backward-Euler integration method, use the MU option:

```
.OPTIONS MU=0
```

The backward-Euler integration method was added to SPICE in a later release, after Nagel[3] had finished his work on the program. For this reason, the MU option was never documented in the original SPICE2 User's Guide.[4] For this reason, many simulation users are not aware of the MU option.

In SPICE, the original trapezoidal algorithm has been modified to include the scaler MU in the formulation. Because of similarities in the formulation of the two algorithms, if the value of MU is 0.5, the formula reduces to the simple trapezoidal integration method. If MU is set to 0, the formula reduces to the backward-Euler integration method. And if MU is set between 0 and 0.5, a weight-averaged value of the two methods is used. In SPICE, the value of MU is set by default to 0.5. Setting the option MU = 0 changes to backward-Euler integration, and setting MU between 0 and 0.5 combines the two integration methods.

Detecting and Correcting Integration Failures

None of the three integration methods in SPICE will always produce an accurate result on all transient circuits. All three integration methods contain different failure mechanisms. Occasionally, during transient analysis, certain voltage and current waveforms will excite one or more of the integration failure mechanisms. Exciting one of these failure mechanisms will introduce error in the simulation output. The amount of the error may range from millivolts and picoamps to tens of volts or amps. It is up to the simulation user to detect numeric integration errors and take corrective actions. SPICE does not have an internal mechanism to detect numeric integration failures and therefore does not flag the erroneous output. *Numeric integration failures must be detected in the simulation output by the user.* Only then can corrective measures be applied.

Trapezoidal integration failures

The trapezoidal integration method is the default method used in SPICE. The trapezoidal method combines good accuracy with low computational requirements. For many circuits, the trapezoidal integration method is the best choice for simulation.

When is the trapezoidal method not a good choice? When the trapezoidal method starts to contribute significant error to the simulation result. There are two well-known failure mechanisms of the trapezoidal integration method that simulation users should learn to recognize. One failure mechanism is trapezoidal oscillation; you have already seen an example of this in Fig. 4.15a. The other failure mechanism is accumulated error. Once the existence of these failure mechanisms is known, both mechanisms can be identified from characteristic features in the simulation output. The characteristic features of these failure mechanisms are easy to identify if you know what to look for in the output. By learning the characteristic features of the failure mechanisms, SPICE users can learn to detect the presence of trapezoidal integration failure and swiftly switch to a more appropriate integration method.

Trapezoidal oscillation.   Trapezoidal oscillation is a well-known, well-documented characteristic of the trapezoidal integration method.[5] Trapezoidal oscillation occurs when the integration step size is too large to follow the curvature of a given function. The result of this failure mechanism is a predicted solution which appears to oscillate around the correct solution. The disk files ch4-17a.cir and ch4-17b.cir contain netlists for the transient simulation of an MOS capacitor. Figure 4.17a shows the simulation result from ch4-17a.cir. The simulation result of Fig. 4.17a appears to oscillate around the correct result predicted by Fig. 4.17b. In simulating the circuit ch4-17a.cir, trapezoidal oscillation introduces large amounts of error in the output. Simulate this circuit with the command:

```
SIM CH4-17A.CIR
```

Your result should match Fig. 4.17a.

To eliminate trapezoidal oscillation from the simulation output, the SPICE user may switch to either Gear or backward-Euler integration. Neither Gear nor backward-Euler integration suffers from the oscillation anomaly. Figure 4.17b shows the simulation result from the circuit ch4-17b.cir. The ch4-17b.cir circuit netlist is identical to ch4-17a.cir except for the addition of the option statement which calls the backward-Euler integration method (.OPTIONS MU=0). The backward-Euler method completely eliminates the oscillation.

Simulate this circuit with the command:

```
SIM CH4-17B.CIR
```

Your result should match Fig. 4.17b.

The ch4-17c.cir circuit netlist is identical to ch4-17a.cir except for the addition of the option statement which calls the Gear-2 integration method (.OPTIONS METHOD=GEAR). The simulation results of ch4-

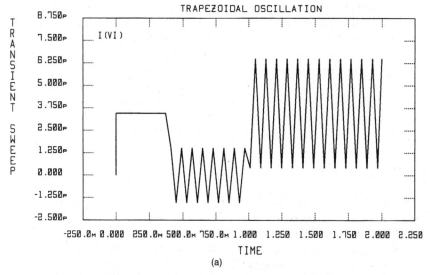

Figure 4.17(*a*)   Transient simulation of an MOS capacitor.   (*Reprinted from* Successfully Simulating Circuits With SPICE. *Used with permission.*)

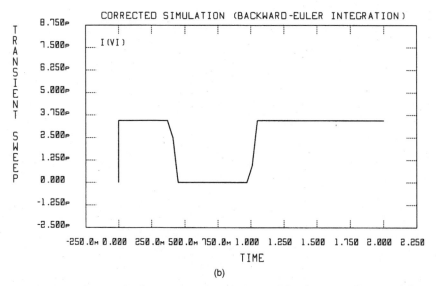

Figure 4.17(*b*)   Corrected transient simulation of an MOS capacitor.   (*Reprinted from* Successfully Simulating Circuits With SPICE. *Used with permission.*)

17c.cir are shown in Fig. 4.17c. Figure 4.17c shows the simulation results of ch4-17c.cir and illustrates a distinct overshoot at both of the transition points. During the simulation of ch4-17c.cir, and on many switching circuits, the Gear integration methods, because of their averaging effect, tend to produce overshoot where circuit voltages and currents change abruptly. Like the oscillation caused by trapezoidal integration, the overshoot is not present in the actual circuit. The overshoot is merely an error introduced by the Gear integration method.

While either the Gear or backward-Euler integration methods could be chosen to eliminate trapezoidal oscillation, for this circuit the backward-Euler method is the best choice. In comparing the simulation results from all three simulations, *and by knowing what the simulated output should be,* the backward-Euler integration method is the best choice for this simulation.

Several SPICE-like simulators do not allow users to switch to Gear or backward-Euler integration. What can the users of these tools do to correct trapezoidal oscillation problems? The answer is to reduce the maximum transient timestep. (Note: Chap. 5 will present a detailed look at how to limit the maximum transient timestep.) Figures 4.14*b*, 4.15*b*, and 4.16*b* demonstrate how reducing the transient step size eliminates, or reduces, both trapezoidal oscillation and Gear overshoot. Decreasing the maximum transient step size is another way of reducing the error and improving the accuracy of all the integration methods in SPICE.

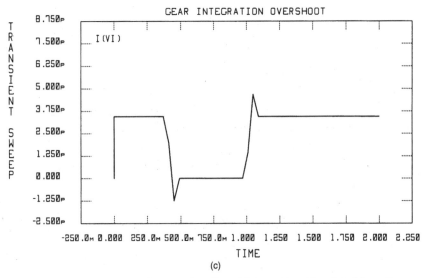

Figure 4.17(c) Transient simulation using Gear-2 integration. (*Reprinted from* Successfully Simulating Circuits With SPICE. *Used with permission.*)

But decreasing the timestep is not always desirable. In Chap. 3, it was shown that decreasing the timestep made the simulator run longer and made simulations more prone to nonconvergence problems. Whenever possible, changing integration methods is the preferred way to correct numeric integration problems.

Trapezoidal oscillation is caused by a failure of the numeric integration method, not by an improper circuit description or device model. At times, as in Fig. 4.17a, trapezoidal oscillation may be large and radically unrealistic; at other times, the oscillation may be very subtle. Not all trapezoidal oscillation is as pronounced as the previous example. Figure 4.18a shows the simulation result of the circuit file ch4-18a.cir. The subtle ripple in the simulation output is the result of trapezoidal oscillation. Figure 4.18b shows the result of the same simulation with the Gear integration method.

Simulate these circuits with the command:

```
SIM CH4-18A.CIR
```

then

```
SIM CH4-18B.CIR
```

Your results should match Figs. 4.18a and b.

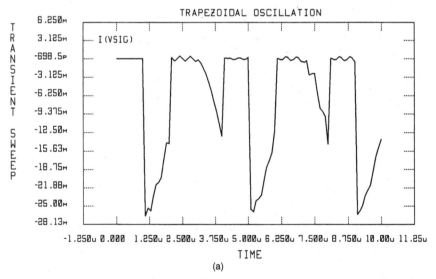

Figure 4.18(a)   Transient simulation with trapezoidal oscillation. (*Reprinted from* Successfully Simulating Circuits With SPICE. *Used with permission.*)

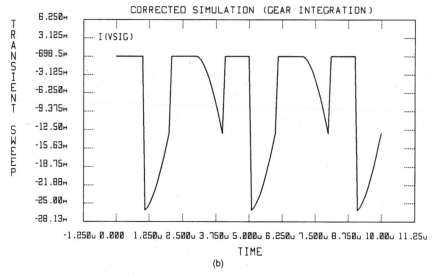

Figure 4.18(*b*)   Corrected transient simulation. (*Reprinted from* Successfully Simulating Circuits With SPICE. *Used with permission.*)

Because SPICE does not have an algorithm to detect trapezoidal oscillation, the program will not warn users of the integration method failure. Trapezoidal oscillation is a phenomenon which must be observed by the user. The clues to look for are unexplained or unexpected oscillation, especially oscillation around what should be the correct result. This naturally leads to the question of how you know whether the oscillation seen in a simulation output is real or merely failure of the trapezoidal algorithm. The best way to differentiate between the two is to rerun the simulation with either the Gear or backward-Euler integration method. With the new integration method selected, the oscillation will disappear if it was caused by a failure of the trapezoidal algorithm. If the circuit response is the cause of the oscillation, all three integration methods—trapezoidal, Gear, and backward-Euler—will reproduce the oscillitory behavior.

Accumulated error.   Accumulated error is a second failure mechanism of the trapezoidal integration method. Accumulated error usually occurs in periodic circuits and long transient simulations. The result of this failure mechanism is usually an unexplainable circuit behavior.

A typical transient simulation may require thousands of timepoint calculations. At each timepoint, the numeric integration algorithm may introduce a finite amount of error in the solution. At some timepoints, the error will overestimate the exact solution, while at other timepoints the error will underestimate the exact solution. Since the error from the previous solution point is carried forward to the next solution point, the overestimation/underestimation errors usually cancel each other over a large number of timepoints. But sometimes the accumulated error tends to increase with each new timepoint. When this occurs, the simulation result deviates from the correct solution. Often the simulation output appears to be completely inaccurate. In many cases, the erroneous output will appear to defy the laws of physics (or electronics).[6]

The circuit file ch4-19a.cir is a listing for the transient behavior of a 555 timer. In the ch4-19a.cir netlist, an external capacitor and several external resistors are connected to the timer. In this configuration, the timer behaves like an asynchronous oscillator. The simulation results are shown in Fig. 4.19a. In this simulation, accumulated error introduces a mathematically stable state which is not present in the real 555 timer. Because the circuit is an asynchronous oscillator, once the integration error accumulates to a point where the circuit is stable, the output remains in that state for the rest of the simulation. Simulate this circuit with the command:

```
SIM CH4-19A.CIR
```

Your result should match Fig. 4.19a.

The circuit file ch4-19b.cir is the same 555 timer circuit, with the addition of the option statement which calls the Gear-2 integration method. The result of this simulation is shown in Fig. 4.19b. Simulate this circuit with the command:

```
SIM CH4-19B.CIR
```

Your result should match Fig. 4.19b.

Accumulated error in trapezoidal integration is usually characterized by an unexplainable simulation behavior. In these cases, switching to a more stable integration method (Gear-2) or reducing the maximum transient timestep minimizes the accumulated error and restores the normal circuit behavior. (Author's note: Experience has shown that accumulated numeric integration error is not the only cause of unexplainable circuit behavior!)

Backward-Euler integration failures

Although the backward-Euler method does not suffer from the numerical oscillation of the trapezoidal method, it may suffer from larger amounts of local truncation error when applied to nonlinear wave-

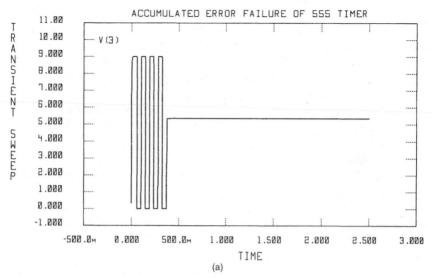

Figure 4.19(a)   Accumulated error simulation failure. (*Reprinted from* Successfully Simulating Circuits With SPICE. *Used with permission.*)

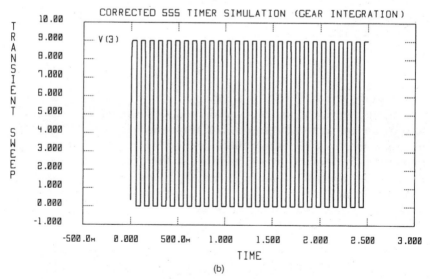

Figure 4.19(b)   Corrected transient simulation. (*Reprinted from* Successfully Simulating Circuits With SPICE. *Used with permission.*)

forms. For sinusoidal and exponential waveforms, the backward-Euler integration method may generate large amounts of error which lead to accumulated-error failures and unexpected simulation results. For these types of circuits, the backward-Euler integration method is not suggested. The backward-Euler is best used on linear and piecewise linear waveforms. The backward-Euler method is a good choice for digital circuits.

The circuit file ch4-20a.cir is a listing for the transient behavior of an ideal resonator circuit. The circuit is composed of a single capacitor and inductor. The transient simulation is started with an initial current flowing through the inductor. Because both of the elements are ideal (no parasitic leakages), the circuit should continue to oscillate for the entire transient simulation. Figure 4.20a shows the simulation result when using the backward-Euler integration method. For sinusoidal and nonlinear waveforms, the backward-Euler method may introduce large amounts of local truncation error in the simulation. On these circuits, the local truncation error quickly accumulates and results in an accumulated-error failure. In the simulation of ch4-20a.cir, the error quickly dampens the ideal oscillation. Simulate this circuit with the command:

```
SIM CH4-20A.CIR
```

Your result should match Fig. 4.20a.

The circuit file ch4-20b.cir is the same oscillator, except for the option which resets the simulator to the trapezoidal integration method. Simulate this circuit with the command:

```
SIM CH4-20B.CIR
```

Your result should match Fig. 4.20b. Figure 4.20b illustrates proper simulation output.

Like trapezoidal accumulated-error failures, accumulated error in backward-Euler integration is characterized by an unexpected or unexplained simulation result. Backward-Euler numeric integration is especially prone to accumulated-error failures when simulating nonlinear or sinusoidal waveforms, and is best used on linear or piecewise linear waveforms.

Gear integration failures

Like backward-Euler integration, Gear integration does not suffer from trapezoidal oscillation, but Gear integration may suffer from large amounts of local truncation error, especially on circuits which contain highly nonlinear or switching waveforms.

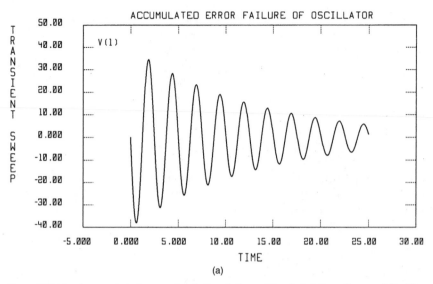

Figure 4.20(a)   Accumulated error simulation failure. (*Reprinted from* Successfully Simulating Circuits With SPICE. *Used with permission.*)

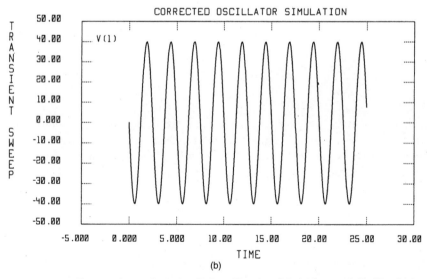

Figure 4.20(b)   Corrected transient simulation. (*Reprinted from* Successfully Simulating Circuits With SPICE. *Used with permission.*)

The Gear integration method uses an averaging technique to predict the next solution point. In theory, this averaging technique should make the algorithm more stable (less accumulated error over a large number of timepoints). But, because the Gear method uses past timepoint values to help predict where the next solution point should be found, for switching waveforms and piecewise linear waveforms, the Gear method may overshoot the correct solution value. (You have already seen an example of this in Figs. 4.16a and 4.17c.) Pronounced overshoot in switching circuits is a characteristic failure of Gear method integration in SPICE.

Many switching circuits exhibit overshoot at the transition points. Once again, simulation users face the dilemma of whether the overshoot seen in their simulation is real or simply a Gear integration error. The best way to determine the true cause of the overshoot is to switch integration methods. Neither trapezoidal nor backward-Euler integration methods generate the overshoot of the Gear integration method. Rerun the simulation with one of the other integration methods. If the overshoot was caused by the mechanics of the circuit, the simulation result should be identical to the Gear integration results. If the overshoot was caused by a Gear-introduced error, the overshoot will disappear.

Suggested Integration Settings

When simulating transient circuits, users must learn to look for numeric integration anomalies. Table 4.2 is a list of suggested integration methods for different types of circuits. Table 4.3 is a summary of numeric integration failure mechanisms and corresponding corrective actions. In Table 4.3, the list of corrective actions includes reducing the maximum transient timestep. Setting the maximum timestep is the subject of the next chapter. Read Chap. 5 before using this technique to correct numeric integration problems. Use these tables to minimize numeric integration failures and correct integration failures if they do occur.

TABLE 4.2   Suggested Integration Methods

Circuit waveforms	Suggested integration method
Sinusoidal (sine wave circuits, oscillators, general amplifiers, power supplies, transmission lines)	Trapezoidal
Linear and piecewise linear (pulse circuits, digital and logic)	Trapezoidal or backward-Euler
Exponential (bipolar and diode amps, RC tree networks)	Trapezoidal or Gear

TABLE 4.3   Common Numeric Integration Failures

Failure mechanism	Corrective action
Trapezoidal oscillation	Switch to backward-Euler or Gear method or reduce maximum timestep
Trapezoidal accumulated error	Switch to Gear method or reduce maximum timestep
Gear overshoot	Switch to backward-Euler or trapezoidal integration or reduce maximum timestep
Gear Accumulated error	Reduce maximum timestep
Backward-Euler accumulated error	Switch to trapezoidal or Gear method or reduce maximum timestep

Numeric Integration in Other Simulators

Hspice

Meta-Software's Hspice contains all three integration methods of standard SPICE, but when Hspice users simulate ch4-17a.cir and ch4-18a.cir they will not see the anomalous behavior seen in the standard SPICE output. The explanation for this has to do more with how Hspice performs transient analysis than how the integration methods are applied.[7]

Standard SPICE and all the other simulators reviewed here, by default, use a timestep control algorithm known as the Local Truncation Timestep Control algorithm. Meta-Software has developed a proprietary timestep control algorithm for transient analysis known as DVDT. The DVDT algorithm improves on several of the deficiencies of the standard timestep control algorithm and as a result tends to suffer from fewer numeric integration problems. (Remember the relationship between step size and numeric integration accuracy.)

A second reason Hspice does not suffer from the same problems is that many of the I-V and C-V device equations have been rewritten and many of the sharp edges of the equations (discontinuities) have been smoothed. Since these same sharp edges initiate trapezoidal oscillation, a reduction in discontinuities reduces the occurrence of trapezoidal oscillation. (Hspice users will notice how the Hspice simulation of circuit ch4-17a.cir produces a very different looking output.)

IS_Spice

Intusoft's IS_Spice is based directly on SPICE 2G.6. IS_Spice offers users the standard choices of trapezoidal, backward-Euler, and Gear integration methods. Trapezoidal integration is the default method in IS_Spice.

Micro-Cap IV

Spectrum Software's Micro-Cap IV is also based on SPICE2G.6 but has been completely rewritten in C. Micro-Cap IV does not support either backward-Euler or the Gear integration methods.[9] If (trapezoidal) integration problems do occur, users must reduce the maximum time-step of the program. Micro-Cap IV users should simulate ch4-17a.cir and ch4-18a.cir to demonstrate the effects of numeric integration failure. Then resimulate the circuits while reducing the maximum time-step until the errors disappear.

Pspice

MicroSim's Pspice uses either trapezoidal or Gear-2 numerical integration. But unlike standard SPICE, Pspice does not offer users an option to manually select the integration method. By default, Pspice starts a transient simulation with trapezoidal numeric integration. As the simulation progresses, the integration routines in Pspice look for the numerical conditions which might induce trapezoidal oscillation. If the conditions for trapezoidal oscillation occur, Pspice automatically changes to Gear-2 integration before the oscillation begins.[10]

Summary

Any of the integration methods in SPICE will produce a finite amount of error at each timepoint in the transient analysis. In many simulations, the error is small. In these cases, the overall accuracy of the simulation result will be within the error tolerances of the program. But when the integration inaccuracies become large, the simulation result will deviate from the correct solution. The amount of deviation may be infinitesimally small or orders of magnitude larger than the original signal! When analyzing simulation results, SPICE users must learn to recognize numeric integration errors. SPICE does not have an algorithm to detect and warn users of numeric integration failure. Be cautious and question your output. Does the simulation generate the result you expect? If not, scrutinize the simulation. Question unexpected behavior. Being aware of integration-caused failures and knowing how to resolve these failures are the watermarks of a true simulation craftsperson.

The previous sections describe the type of errors that SPICE numeric integration methods are most likely to produce. These are the types of erroneous behavior SPICE users must learn to look for. SPICE users should never run a simulation without some reasonable expectation of what the output *should* be. This statement is important, because without a reasonable expectation of the output, detecting numeric integration failures may be impossible. Because SPICE and almost all of the SPICE-like simulators use trapezoidal integration as the default integration method, you should begin looking for trapezoidal integration failures first. The two integration failures associated with trapezoidal integration are trapezoidal oscillation and accumulated error. Of these two failure mechanisms, trapezoidal oscillation is the most common. Figures 4.17a and 4.18a are illustrations of trapezoidal oscillation. Users should also look for signs of accumulated integration error. Figures 4.19a and 4.20a are illustrations of accumulated-error failures.

When you switch to the Gear integration method, watch for unexpected overshoot in switching circuits, and watch for accumulated error. When you switch to the backward-Euler integration method, watch for accumulated error failures.

In SPICE, the numeric integration algorithms routinely produce stunningly accurate, highly reliable results during transient simulation. But occasionally, the numeric integration routines will generate erroneous results. Since SPICE does not automatically detect numeric integration errors, the responsibility of detecting and correcting these errors falls on the user. Most numeric integration failures can be readily identified if users know to look for them, and all numeric integration failures can be fixed. Correcting the integration failure may involve switching algorithms or reducing the maximum internal timestep. Becoming a knowledgeable SPICE user involves much more than simply learning to create input netlists. It also involves knowing the strengths and weaknesses of SPICE, and knowing how to compensate for the weaker facets of the program.

References

1. Leonard Euler, 1709–1787.
2. C. W. Gear, *Numerical Integration of Stiff Ordinary Equations,* University of Illinois at Urbana-Champaign, Report 221, January 20, 1967.
3. L. W. Nagel, *SPICE2: A Computer Program to Simulate Semiconductor Circuits,* Electronics Research Laboratory Rep. No. ERL-M520, University of California, Berkeley, 1975.
4. *SPICE Version 2G.6 User's Guide.*
5. Leon O. Chua & P. M. Lin, *Computer-Aided Analysis of Electronic Circuits: Algorithms and Computational Techniques,* Prentice-Hall, 1975.
6. Author's personal experience.
7. Kim Hailey, Meta-Software, Inc., private communications.
8. Charles Hymowitz, Intusoft, private communications.
9. Andrew Thompson, Spectrum Software, private communications.
10. Graham Bell, MicroSim Corporation, private communications.

5

Timestep Control

The timestep control algorithms determine the timepoints where SPICE solves the circuit equations. The selection of timepoints relates directly to the accuracy of the numeric integration routines and indirectly to the convergence properties of the simulator. For these reasons, the timestep control algorithms of SPICE are the most important routines of the transient simulation. But like so many other routines in SPICE, the timestep control algorithms are not foolproof. Accurate, high-quality transient simulations require a precise tailoring of the timestep control routines.

This chapter will present a look at the two timestep control algorithms in SPICE. Like numeric integration, each algorithm has advantages and disadvantages, each algorithm will work well with certain types of circuits, and each algorithm will suffer from a unique set of problems. Knowing which timestep control algorithm to use on a specific circuit and knowing the limitations of each algorithm is the next lesson in becoming a true simulation craftsperson.

Timestep Control in the Early Years

In the late 1960s and early 1970s when CANCER and SPICE1 were being used to simulate circuits, the transient command:

```
.TRAN 1NS 100NS
```

forced the simulator to solve the circuit equations at each 1nS time interval. This is known as constant timestep control. With this command, both CANCER and SPICE1 solved the circuit equations and printed the output results every 1nS during the transient simulation. Figure 5.1a illustrates a constant timestep control simulation. For these early simulators, the print interval and the internal timepoints were identical.

But in sweeping the timestep at a constant rate, two problems were soon observed. First, circuits with large, rapid voltage and current transitions often failed to converge at the switch point. (This is due to the same nonconvergence mechanism which causes the DC sweep analysis to fail around voltage transitions. See Chap. 3 for a thorough discussion of why this occurs.) Only by significantly reducing the step size would the simulation successfully cross the voltage transition. But small timesteps generate extremely long simulation times and may aggravate nonconvergence problems. The second problem with constant timestep control relates to the numeric integration algorithms. With a constant step size, numeric integration algorithms produce the most accurate result when the circuit is in a stable state, and the integration routines produce the least accurate solution when the circuit is in a transition state. *With a constant step size, the error introduced by the numeric integration algorithms is proportional to the transition rate of circuit voltages and currents.* For circuit simulation, constant timestep control forces users to select between a slow (long run time), accurate simulation and a fast, inaccurate one.

Improving the Timestep Control Algorithm

The introduction of SPICE2 brought two different forms of dynamic timestep control to circuit simulation. Both of the timestep control algorithms decrease the step size when the circuit approaches a volt-

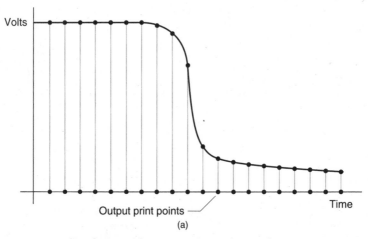

(a)

Figure 5.1(a)   Simulation with constant timestep control.

age or current transition and then increase the step size when the circuit begins to stabilize again. These dynamic timestep control algorithms are substantially better than simple constant timestep control and significantly increase both the accuracy and speed of a transient simulation.

But, like the numeric integration algorithms, neither of the timestep control algorithms are completely foolproof. Both timestep control routines have problems on certain types of circuits, and these problems may introduce significant error in the simulation output. Even with completely accurate device models, *failure mechanisms within the timestep control algorithms may introduce significant error into the simulation result.* A knowledgeable simulation user is aware of these possible problems and, when questionable simulation results arise, knows the corrective action to employ.

An Overview of Dynamic Timestep Control

The two timestep control algorithms in SPICE both have basically the same purpose. The timestep control algorithms must alter the step size so, during rapid voltage transitions, the timestep is decreased to avoid nonconvergence and maintain accuracy through the transition region, and, during periods of little or no circuit activity, the timestep is increased to speed the simulation to a conclusion. But there are subtle differences between the two timestep control algorithms which may amount to significantly different simulation results.

The dynamic timestep control algorithms significantly changed the way SPICE computes transient timepoints, but the transient command did not change when the timestep control routines changed. This has lead to a slightly confusing situation. The transient command

```
.TRAN 1NS 100NS
```

still seems to imply that SPICE solves the transient simulation at evenly spaced 1nS intervals. This command is the same command the earlier version of CANCER and SPICE1 used with their constant timestep control routines. The dynamic timestep control routines do not solve the circuit equations at evenly spaced intervals. The timestep control routines perform the task of deciding where the circuit equations should be solved. As the transient analysis begins, SPICE uses Eq. 5.1 to compute the first transient timepoint. After the program computes the first timepoint, the timestep control algorithms assume the responsibility for increasing or decreasing the size of the timestep.

$$STEP(1) = \frac{\text{transient duration}}{50} \qquad (5.1)$$

As SPICE continues the transient simulation, the timestep control algorithms monitor and adjust the size of the timestep. At each computed timepoint, SPICE saves the solution values in memory. At the end of the transient simulation, SPICE uses a simple linear interpolation routine and the data saved from the transient simulation to generate the evenly spaced 1nS output result. Figure 5.1*b* illustrates a dynamic timestep control simulation. In most simulations, the number of timepoints SPICE solves during the transient analysis is much higher than the number of output points. The 1nS print interval has very little influence on the selection of timepoint values.

Both timestep control algorithms found in SPICE use three sensors to control the size of the timestep. The sensors monitor the rate of change of circuit voltages and currents, nonconvergence timepoints, and source breakpoints. Of these, the first, the rate-of-change sensor, is the predominant mechanism which influences the selection of timesteps. The others modify the timestep predicted by the rate-of-change sensor. In addition to the control sensors, the timestep control algorithms also have upper and lower limits on the size of the timestep. All these mechanisms work together to predict the best timestep for the simulation.

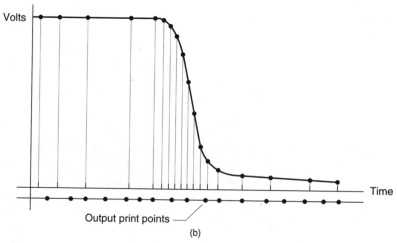

(b)

Figure 5.1(*b*)   Simulation with dynamic timestep control.

Dynamic timestep control increases both simulation speed, by sampling fewer timepoints, and simulation accuracy, by sampling more points where they are needed. SPICE offers users two types of dynamic timestep control.

When using dynamic timestep control, the values printed in the output file are interpolated from the two closest solution points. The print interval has little influence on the selection of solution timepoints.

Timestep control and circuit activity

Both timestep control algorithms have a method of measuring the rate of change of the circuit voltages and currents, and this measurement will be used to increase or decrease the current timestep. The timestep is decreased during voltage and current transitions and increased during periods of lesser circuit activity. Each of the timestep control algorithms uses different measurements to determine the circuit's rate of change. One of the algorithms monitors the number of iterations at each new solution point. During rapid circuit transitions, the change in voltage and current levels requires more Newton-Raphson iterations to achieve convergence. As the number of iterations increases, the timestep is decreased. During periods of latency, far fewer iterations are required because the circuit voltages are not changing. Under these conditions, the timestep can be increased. As the analysis progresses, the step size is adjusted, based on the number of iterations used at the past timepoint. The other algorithm uses a routine which monitors the amount of error the numeric integration routines are producing. During periods of little or no circuit activity, the numeric integration routines are extremely accurate and produce very little error. Because the error is low, the timestep control algorithm increases the step size. As the circuit activity increases, the error increases, and this forces the timestep control algorithm to decrease the timestep. Throughout the entire transient analysis, the rate-of-change sensors monitor the circuit activity and make appropriate adjustments to the timestep.

Timestep control
and nonconvergent timepoints

Both timestep control algorithms measure the number of iterations required to converge on a solution timepoint. *If a given timepoint ever fails to converge to a solution within ITL4 iterations, both timestep control algorithms will discard the failed timepoint, reduce the previous timestep to ⅛ of the original value and then reattempt the solution at a timepoint closer to the last convergent timepoint.* Nagel et al.[1] determined that when the transient simulation failed to converge within a set number of iterations, the failure could usually be attributed to a fast circuit transition. Because of this, Nagel reduced the timestep to ⅛ of the original value and reattempted a solution at the new timepoint. Figure 5.2 illustrates how SPICE recalculates a new timepoint after failing to converge.

The ability to discard nonconvergent timepoints and reattempt a new solution is a significant improvement over both the DC operating

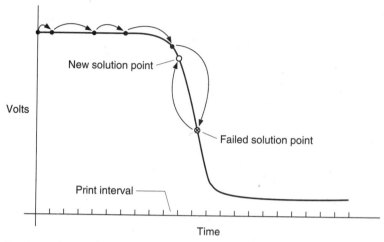

Figure 5.2   Attempting a new solution point after a nonconvergent timepoint.

point and the DC sweep solution algorithms. Only during transient analysis will the simulator keep searching for a solution after a non-convergent timepoint has been found. SPICE will continue to search for a new solution point by reducing the timestep over and over until either the simulator converges or the size of the timestep is reduced below an allowed minimum. Only when the latter occurs does SPICE halt the solution algorithm and print the "internal timestep too small" warning message.

Timestep control and source breakpoint adjustment

Both pulse and PWL (piecewise linear) circuit sources generate pulse waveforms with sharp leading and trailing edges. Often, a change in the state of the source, such as a leading edge or trailing edge, is a pre-cursor to circuit transition. Because of this, SPICE generates a break-point table before the start of the transient simulation. The table contains the timepoints corresponding to the beginning and end of each rising or falling edge for pulse or PWL sources. During transient analysis, SPICE forces the timestep control algorithm to solve the cir-cuit at each breakpoint, and automatically reduces the timestep fol-lowing a breakpoint calculation. Forcing the simulator to solve the circuit at the source breakpoints greatly increases the accuracy of the transient output. Figure 5.3 illustrates the breakpoints of a pulse volt-age source.

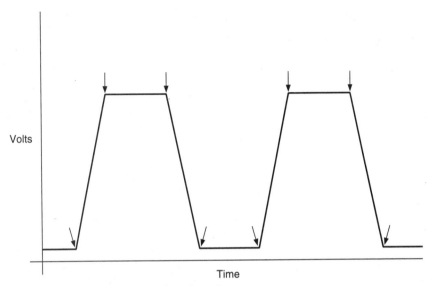

Figure 5.3   Source breakpoints of a pulse voltage source.

Minimum and maximum step sizes

Each of the timestep control algorithms has an upper and lower limit on the size of the timestep. During long periods of latency, the timestep control algorithm continues to try to increase the timestep. The timestep will be increased until it reaches the maximum allowed step size. Even with no circuit activity, the timestep will not exceed this maximum limit. If a solution point fails to converge, SPICE will reduce the size of the timestep by 8 and reattempt the solution. If the new timepoint fails to converge, the timestep will be reduced by 8 again. This process repeats until a convergence solution is found or until the step size reaches or becomes less than the minimum. If the step size is reduced below the allowed minimum, SPICE prints the "internal timestep too small" error message and aborts the simulation.

For both timestep control algorithms, the mechanisms just described determine the size of the timestep and where SPICE tries to solve the circuit equations. The rate-of-change sensors select a timestep which is appropriate for the circuit. If the new timepoint fails to converge, the timestep is reduced by a factor of 8 and the solution reattempted. If the timestep is close to a source breakpoint, the step size is adjusted to align with the breakpoint. An inactive period of the simulation may increase the timestep to the maximum allowed limit, or a nonconvergence problem may reduce the timestep below the minimum allowed limit and abort the simulation. These are the mechanisms which determine the step size during transient analysis.

Iteration-Count Timestep Control

The first of the two timestep control algorithms is the Iteration-Count method. The name comes from the way SPICE monitors the circuit activity.

The Iteration-Count timestep control algorithm monitors the number of iterations at each timepoint during the transient simulation. With the circuit in a stable condition, the number of iterations required to solve the circuit equations at each new timepoint is usually small. When the number of iterations becomes smaller than the lower threshold (ITL3), the timestep control algorithm doubles the size of the timestep before calculating the next timepoint. A low number of iterations indicates the circuit is relatively inactive and therefore a larger timestep can be used.

When the circuit begins to enter a transition or the circuit activity starts to increase, the number of iterations at each timepoint increases. When the number of iterations exceeds the upper limit (ITL4), the timestep control algorithm discards the current timepoint, cuts the timestep to ⅛ of the present value, and reattempts the solution with the smaller timestep.

If the circuit enters a latent phase and the number of iterations is lower than ITL3, the timestep control routine will try to increase the timestep. If the circuit remains in a latent stage, the timestep control will continue to increase the step size until the size of the timestep reaches its maximum limit. With the Iteration-Count method, the timestep is limited to become no larger than the print interval specified on the transient statement. The print interval determines the maximum timestep for the Iteration-Count method.

If the current timepoint fails to converge, the timestep will be reduced to ⅛ of the previous value and the solution reattempted at the new timepoint. If the new timepoint fails, the timestep will again be cut. This process will continue until either a convergent timepoint is found, or the step size is reduced below the allowed minimum. For the Iteration-Count timestep control algorithm, the minimum allowed step size is TSTOP/50e9, where TSTOP is the duration of the transient analysis.

The Iteration-Count timestep control algorithm adjusts the timestep by using an upper and lower threshold. When the number of iterations is greater than the upper threshold, ITL4, the timestep is reduced by a factor of 8. When the number of iterations is less than the lower threshold, ITL3, the timestep is increased by a factor of 2. The Iteration-Count algorithm adjusts the timestep to keep the number of iterations at each timepoint between these two thresholds.

Regardless of the circuit activity (or inactivity), the timestep is never allowed to become larger than the print interval. If the timestep is reduced below the allowed minimum, SPICE prints the "internal timestep too small" error message and aborts the simulation.

Both ITL3 and ITL4 are SPICE options and may be adjusted by the user. ITL3 is set by default to 4 iterations, and ITL4 is set by default to 10 iterations. The Iteration-Count timestep control algorithm is selected by setting the option LVLTIM to 1.

Local Truncation Error Timestep Control

The second type of timestep control is the Local Truncation Error timestep control algorithm. The Local Truncation Error method is the default method used by SPICE and most of the SPICE-like simulators. The Local Truncation Error routine uses a formula which predicts the magnitude of the error computed in the numeric integration calculations of the previous timepoint. The Local Truncation Error timestep control algorithm uses the amount of error being generated in the numeric integration routines to adjust the timestep. Equation 5.2 illustrates the equation relating the calculated local truncation error and the timestep. The expression LTE in Eq. 5.2 is the computed local truncation error. As the truncation error increases, the timestep decreases; as the truncation error becomes smaller, the timestep grows. This algorithm maintains the timestep so the amount of error in the numeric integration routine is maintained within an allowed limit.

$$\text{STEP(T + 1)} = \sqrt{\frac{\text{TRTOL*(RELTOL*} \, |\, I_c(T)\, |\, + \text{ABSTOL})}{\text{MAX(ABSTOL, LTE(T))}}} \quad (5.2)$$

In Eq. 5.2, the computed local truncation error is scaled by the options ABSTOL, RELTOL, and TRTOL. ABSTOL and RELTOL are the convergence error tolerance options presented in Chap. 3; TRTOL will be discussed later in this chapter. Equation 5.2 re-illustrates the importance of setting the error tolerance parameters to appropriate levels for the voltages and currents in your circuit. Inappropriate settings for ABSTOL and RELTOL may force SPICE to use an unusually small timestep during transient simulation. While a small timestep will help ensure accuracy of the simulation, an excessively small timestep will cause long run times and may aggravate nonconvergence problems.

In addition to the local truncation error rate-of-change sensor, the Local Truncation Error timestep control algorithm also uses ITL4 as a limit on the number of iterations allowed at a given timepoint. If the number of iterations at a timepoint is greater than ITL4, the timestep control algorithm cuts the timestep to ⅛ of the original value and reattempts the solution at the new timepoint.

As the circuit activity decreases, the local truncation error intro-duced by the numeric integration routines decreases. As the local trun-cation error decreases, the timestep control routine tries to increase the timestep. If the circuit remains stable, the timestep control routine continues to increase the step size until the step size reaches the max-imum allowed limit. For the Local Truncation Error timestep control routine, the maximum allowed timestep is TSTOP/50, where TSTOP is the transient duration.

If the current timepoint fails to converge, the timestep will be reduced to ⅛ of the previous value and the solution reattempted at the new timepoint. If the new timepoint fails, the timestep will again be cut. This process will continue until either a convergent timepoint is found, or the step size is reduced below the allowed minimum. For the Local Truncation Error timestep control algorithm, the minimum allowed step size is TSTOP/50e9, where TSTOP is the total transient duration. *Both of the timestep control algorithms in SPICE use the same minimum timestep limit.*

The Local Truncation Error timestep control algorithm is the default algorithm in SPICE. The Local Truncation Error method is selected with the option LVLTIM=2.

A Comparison of Timestep Control Methods

While both timestep control algorithms are good, the Local Truncation Error algorithm *should* produce a more accurate transient analysis because of the close coupling between the numeric integration trunca-tion error and the size of the timestep. The Iteration-Count method has no measure of the numeric integration error and therefore no means to adjust the timestep with this error.

But the Local Truncation Error algorithm will only provide a supe-rior result if the estimate for local truncation error is accurate. The for-mula which SPICE uses to calculate the local truncation error is known as the Third Divided Difference formula.[1] Experimentally, Nagel et al. noticed that the local truncation error predicted from the Third Divided Difference equation was generally about a factor of 7 over what the actual error should be. For this reason, Nagel added the option TRTOL to the Local Truncation Error timestep formula (Eq. 5.2). TRTOL is set by default to 7. Even with this correction factor, for certain types of circuits the estimate for local truncation error is extremely poor. For sinusoidal circuits and inductive circuits, the Third Divided Difference formula and Eq. 5.2 lead to an extremely poor estimate of timestep. Because of this, both of these types of circuits may produce better results when simulated with the Iteration-Count timestep control algorithm.

Even when the estimate of local truncation error is accurate, a second failure mechanism exists in the Local Truncation Error timestep control algorithm. The Local Truncation Error timestep control algorithm limits the maximum timestep to TSTOP/50, whereas the Iteration-Count timestep control algorithm limits the maximum timestep to TSTEP.

TSTEP is the print interval and TSTOP is the transient duration. For circuits with more than 50 solution points, the Local Truncation Error timestep control algorithm allows the timestep to become much larger than the print interval. For simulations requiring hundreds or thousands of solution points, the Local Truncation Error timestep control algorithm may use a step size much larger than the print interval. When this occurs, the simulation result is undersampled or *aliased* and the output appears choppy and distorted.

Timestep Control Failures

Figure 5.4 is the schematic of a 200-MHz sinusoidal oscillator. The disk file ch5-5a.cir contains the listing of the circuit. Simulate this circuit in RSPICE with the command:

```
SIM CH5-5A.CIR
```

Your results should match Fig. 5.5*a*.

In this simulation, the default timestep control algorithm (Local Truncation Error) allows the step size to become too large to accurately simulate the circuit. The local truncation error measurement for sinusoidal waveforms is often poor and results in a poor estimate of the step size. The Local Truncation Error timestep control algorithm allows the timestep to grow up to one-fiftieth of the total transient duration (in this case 4nS). For a 200-MHz signal, 4nS is much larger than the worst-case Nyquist criterion of 2.5nS. The 200-MHz signal is undersampled!

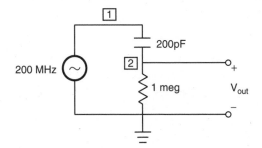

Figure 5.4 Circuit schematic for circuit file ch5-5.cir.

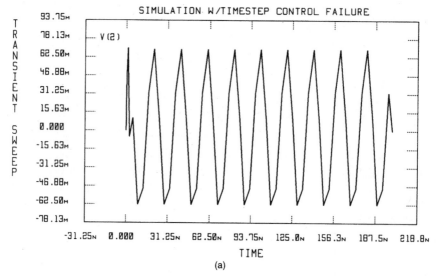

Figure 5.5(a)   (*Reprinted from* Successfully Simulating Circuits With SPICE. *Used with permission.*)

Now simulate the circuit file ch5-5b.cir with the command:

```
SIM CH5-5B.CIR
```

Your results should match Fig. 5.5*b*. In this simulation, the .OPTION LVLTIM=1 directs SPICE to use the Iteration-Count timestep control algorithm. Notice the improvement in the simulation results. The Iteration-Count timestep control algorithm limits the timestep to no more than the print interval TSTEP (in this case .2nS) and, for this circuit, eliminates the distortion caused by undersampling.

Limiting the maximum step size

Although the *default* limit for the maximum step size of the Local Truncation Error timestep control algorithm is TSTOP/50, the maximum step size may be manually limited by the user.

```
.TRAN TSTEP TSTOP <TSTART> <HMAX>
```

The maximum step size of either timestep control algorithm may be limited manually with the HMAX parameter of the transient statement.[2] Both timestep control algorithms use either the *default maximum timestep* or HMAX (whichever is smaller) as the maximum transient step size.

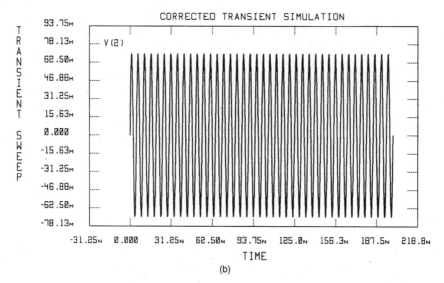

Figure 5.5(b)   The Local Truncation Error timestep control algorithm allows the timestep to become much larger than the print interval. Switching to the Iteration-Count method or manually limiting the HMAX parameter corrects the problem. (*Reprinted from* Successfully Simulating Circuits With SPICE. *Used with permission.*)

Now simulate the circuit file ch5-5c.cir with the command:

```
SIM CH5-5C.CIR
```

Your results should again match Fig. 5.5b. In this simulation, the (default) Local Truncation Error timestep control algorithm's maximum step size is limited to the print interval (HMAX is set to match TSTEP) with the command:

```
.TRAN .2NS 200NS 0 .2NS
```

Again notice the improvement over the initial simulation result. Also notice that if HMAX is to be set, a numeric value must fill the TSTART position!

For those who are not familiar with the TSTART parameter, TSTART specifies the transient simulation time where SPICE begins *printing* the output results. TSTART is not where SPICE begins the simulation. (Transient analysis always begins at time T=0.)

Simulate the disk file ch5-5.cir with the command:

```
RSPICE CH5-5.CIR
```

Then graph the result with the command:

```
RGRAPH CH5-5.OUT CH5-6.MAC /MULTI /MARKERS
```

Your result should match Fig. 5.6. In this simulation, the oscillator circuit is simulated *with and without the HMAX limit.* The two results are superimposed for comparison. Notice how the simulation without the step size limit undersamples the corrected output.

For many simulations, using either the Local Truncation Error timestep control with HMAX set to the print interval or using Iteration-Count will produce nearly identical results. Some simulations will require that HMAX be set to something smaller than the print interval; some simulations will produce an accurate result even with HMAX set larger than the print interval. A good rule of thumb for most simulations is to set HMAX equal to the print interval. As an error-prevention measure, always set the HMAX parameter on the transient statement to help minimize timestep control problems.

Limitations of the breakpoint adjustment

In addition to adjusting the step size as the circuit activity changes, and limiting the step size to a maximum value, SPICE also aligns the solution timepoints with the breakpoints of the sources. As the transient simulation begins to approach a breakpoint, the timestep control algo-

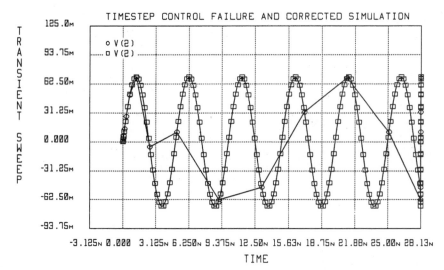

Figure 5.6 Timestep control failure and corrected simulation. (*Reprinted from* Successfully Simulating Circuits With SPICE. *Used with permission.*)

rithms adjust the timestep so the next solution point falls on the next breakpoint. To further enhance the accuracy just after the breakpoint, the timestep control algorithms automatically reduce the timesteps just after the breakpoint in anticipation of a voltage or current transition. For many circuits, timestep control with breakpoint adjustment significantly improves the accuracy of the simulation result.

While the breakpoint adjustment works well on many circuits, when the step size becomes too large the timestep control algorithm may miss one or more source breakpoints. Figure 5.7 is the schematic of an SR flip-flop and test generators. The disk circuit ch5-8a.cir contains the listing of the circuit. The pulse generator uses short pulses to set and reset the flip-flop. Simulate the circuit with the command:

```
SIM CH5-8A.CIR
```

Your results should match Fig. 5.8a. The pulse being used to set and reset the latch is illustrated in Fig. 5.9.

In this simulation, the breakpoints of the short pulses are skipped by the timestep control algorithm! Figure 5.9 shows that the set and reset pulses last only 750nS from start of the rising edge to the end of the falling edge. Because of the short duration of the set and reset pulse and the large timestep, the timestep control algorithm misses the breakpoints and completely skips over the set and reset pulses. As a result, the flip-flop never receives the set or reset signal.

This is a fairly common occurrence in simulations which contain short pulses. Short is a relative term. A more general description is: Any circuit with a voltage or current pulse with a total duration of less than $\frac{1}{100}$ of the simulation's maximum timestep may be subject to problems with the breakpoint adjustment during transient simulation.

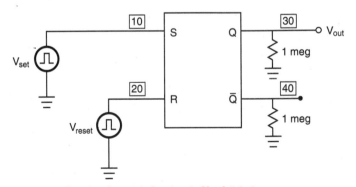

Figure 5.7 Circuit schematic for circuit file ch5-8.cir.

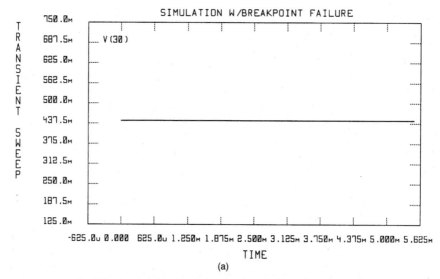

Figure 5.8(a) In this simulation, the timestep control algorithm skipped several source breakpoints. (*Reprinted from* Successfully Simulating Circuits With SPICE. *Used with permission.*)

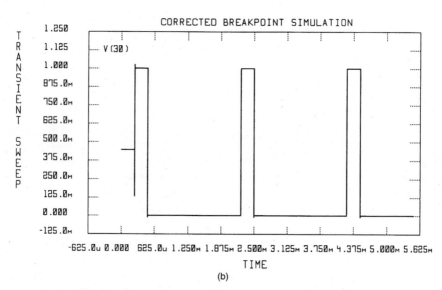

Figure 5.8(b) The breakpoint detection algorithm skipped the short-duration pulses. For these types of circuits, an accurate simulation may require limiting the HMAX parameter. (*Reprinted from* Successfully Simulating Circuits With SPICE. *Used with permission.*)

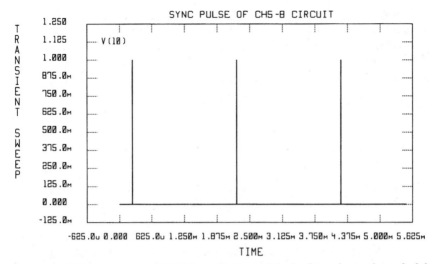

Figure 5.9   The sync pulse is 750nS from the start of the leading edge to the end of the falling edge. With a transient duration in the microsecond range, this small sync pulse could easily be missed by the breakpoint detection algorithm. For these types of circuits, setting the maximum timestep is crucial.

For circuits which use short pulses, set the maximum timestep to no more than 50 times the minimum pulse duration. This will help ensure that the breakpoint adjustment algorithm will detect the presence of the source breakpoint and adjust the timestep accordingly.

In the case of the ch5-8 circuits, the pulse duration is 750nS (250nS rise + 250nS fall + 250nS duration). The above rule states the maximum timestep for this circuit should be:

```
50 * 750nS = 37.5uS
```

The circuit file ch5-8b.cir is identical to the original circuit file except for the addition of the HMAX parameter on the transient statement. In the transient statement, HMAX is set to 37.5uS. Simulate the ch5-8b.cir circuit file with the command:

```
SIM CH5-8B.CIR
```

Your output should match Figure 5.8*b*.

An astute reader may also recognize that long pulse waveforms with short rise and fall times might also be subject to breakpoint problems. While this is true, pulses which have short rise and fall times but a long pulse duration create less problems than short pulse durations. Imagine a long pulse with a very short rise time. Although SPICE may miss the rising edge source breakpoint, if the pulse duration is long the next timepoint will occur with source at its pulsed value. So while the timing might be off slightly, the simulation does not entirely skip the pulse. For those circuits where the timing of the rising and falling edge is critical, set the maximum timestep to no more than 50 * (rise time) or 50 * (fall time) to ensure the breakpoints of the rising and falling edges are found.

Timestep Control Settings for Typical Circuits

The previous circuit examples illustrate the common problems with the SPICE timestep control algorithms. Table 5.1 summarizes the differences between the timestep control algorithms in SPICE. Table 5.2 prescribes a general set of guidelines which should be used to select the timestep control algorithm and the maximum step size. Using Table 5.2 to select a timestep control algorithm and maximum timestep will eliminate most of the problems caused by the timestep control algorithms.

Timestep Control and Oscillator Circuits

Oscillator circuits are a special class of circuits that present unique problems for both timestep control algorithms. This is especially true in the simulation of the oscillator during the start-up period.

Figure 5.10 is a simple Colpits oscillator. The disk file ch5-11a.cir is a listing of the oscillator. Simulate this circuit with the command:

```
SIM CH5-11A.CIR
```

Your results should match Fig. 5.11a.

TABLE 5.1   SPICE Timestep Control Algorithm Summary

Characteristic	Iteration count	Local truncation error
Initial timestep	TSTOP/50	TSTOP/50
Minimum timestep	TSTOP/50e9	TSTOP/50e9
Maximum timestep (without HMAX)	TSTEP	TSTOP/50*
Maximum timestep (with HMAX)	MIN(HMAX,TSTEP)	MIN(HMAX,TSTOP/50)*
.OPTION ITL3=X (default=4)	Iteration limit to increase timestep	Not used
.OPTION ITL4=X (default=10)	Iteration limit to reduce timestep	Iteration limit to reduce timestep

*If circuit contains no energy storage elements (capacitors or inductors), the maximum timestep for LTE without HMAX is TSTEP, and the maximum timestep with HMAX is MIN(HMAX,TSTEP).

TABLE 5.2   General Timestep Control Algorithm Settings with Maximum Timestep Recommendation

Nature of circuit	Timestep control algorithm
General circuits	Local Truncation Error w/HMAX = TSTEP
Sinusoidal with frequency = FREQ	Local Truncation Error w/HMAX = TSTEP or w/HMAX = 1/(8*FREQ)
High inductive content	Iteration-Count or Local Truncation Error w/HMAX = TSTEP
Pulse-driven circuit	Iteration-Count w/HMAX = 50 * Shortest Pulse Duration or Local Truncation Error w/HMAX = 50 * Shortest Pulse Duration

In this circuit, SPICE fails to predict the start-up of the oscillator. The reason SPICE fails to predict the oscillation relates to the operation of the timestep control algorithm.

Most real oscillators start oscillating because of minute voltage and current surges (noise) within the circuit. SPICE also generates minute amounts of voltage and current noise in the form of round-off error. The error tolerance options of the program (RELTOL, VNTOL, and ABSTOL) determine the magnitude of the round-off error noise. So why doesn't the oscillator start? The oscillator fails to start because the voltage surges *for the size of the timestep being used* were not large enough to unsettle the circuit.

Assume the voltage error tolerance of the program is set to 1uV (.OPTION VNTOL=1uV). The minimum amount of numerical noise at each timepoint is ±1uV. The voltage surge at each timepoint is delta_V/delta_T or 1uV/timestep. As the transient simulation begins, SPICE computes an output voltage very close to zero volts. Oscillation

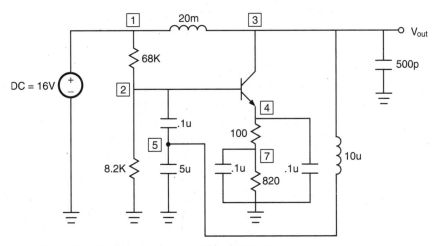

Figure 5.10   Circuit schematic for circuit file ch5-11.cir.

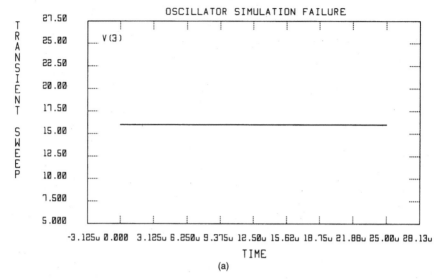

Figure 5.11(a)   (*Reprinted from* Successfully Simulating Circuits With SPICE. *Used with permission.*)

has not yet begun. As the transient simulation continues, the step size grows quickly because the timestep control algorithm senses no voltage change at the output. As the step size continues to grow, the value of the numerical noise (dV/dt) decreases because the 1uV voltage is now being divided by larger and larger timesteps. The cycle continues. The timestep control algorithm increases the timestep because oscillation has not started, and oscillation does not start because the amount of numerical noise decreases with each larger timestep.

Often, after just a few transient timepoints, the timestep is too large to produce a sufficient dV/dt to start the oscillator and, because of this, the oscillator never oscillates.

To correct this situation, when simulating oscillators during start-up, the maximum timestep must be set to a small value, small enough so the numerical noise can produce a sufficient dV/dt to start the oscillator. A good rule of thumb for oscillators is to *set the maximum timestep to ⅛ of the period of oscillation.*

The disk file ch5-11b.cir is the same Colpits oscillator with the maximum timestep (HMAX) set at 1/(8 * FREQ), where *FREQ* is the natural frequency of oscillation. Simulate this circuit with the command:

```
SIM CH5-11B.CIR
```

Your results should match Fig. 5.11*b*.

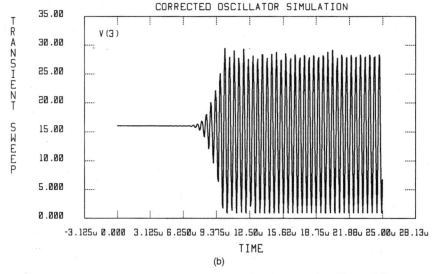

Figure 5.11(*b*)   Simulating oscillators may cause timestep control problems. Often, accurate simulation of these circuits requires limiting the HMAX parameter. (*Reprinted from* Successfully Simulating Circuits With SPICE. *Used with permission.*)

Setting the maximum timestep allows this simulation to correctly predict the behavior of the oscillator. The maximum timestep required to start oscillation will vary from circuit to circuit, but the ⅛th rule is generally a good starting value.

Even after setting the maximum timestep, some oscillator circuits will still fail to start. For these circuits, ramping the power supply from zero up to full power often results in successful start-up. Figure 5.12 illustrates a CMOS ring oscillator. Simulate the circuit disk circuit ch5-13a.cir with the command:

```
SIM CH5-13A.CIR
```

Your results should match Fig. 5.13*a*.

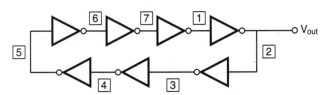

Figure 5.12   Circuit schematic for circuit file ch5-13.cir.

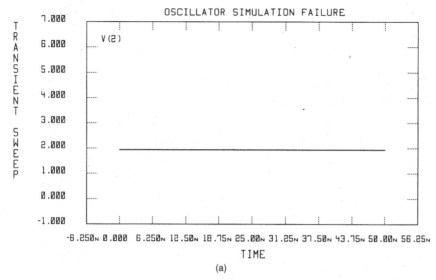

Figure 5.13(a)   (*Reprinted from* Successfully Simulating Circuits With SPICE. *Used with permission.*)

The ch5-13a.cir circuit file is an inverter oscillator (ring oscillator). The circuit uses a series combination of seven CMOS inverters. Notice the maximum timestep (HMAX) has manually been set to ⅙ the period of oscillation in the circuit file and yet the oscillator still fails to oscillate. In simulating ch5-13a.cir, the circuit power supply is set to 5VDC. In the circuit file ch5-13b.cir, the power supply will be ramped with the aid of a PWL source description. In this simulation, the power supply transits from zero to 5VDC much like applying power to a dead circuit. Simulate the ch5-13b.cir circuit with the command:

```
SIM CH5-13B.CIR
```

Your results should match Fig. 5.13*b*.

On occasion, limiting the maximum timestep and ramping the power supply will still fail to start some stubborn oscillators. One sure way of starting the most stubborn oscillator circuit is to use the .IC statement to initialize one of the nodes within the oscillator's loop. But this is an artificial means of initiating oscillation and will not represent the real start-up process of the oscillator.

Oscillators by their very nature can lead to timestep control problems and simulation failure. Setting the maximum timestep to an appropriate value, ramping the power supplies, or initializing nodes

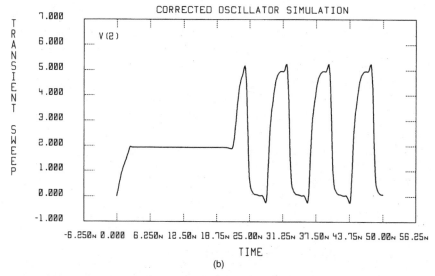

Figure 5.13(b)   In addition to limiting the HMAX parameter, some oscillators will require ramping the power supplies to simulate the onset of oscillation. (*Reprinted from* Successfully Simulating Circuits With SPICE. *Used with permission.*)

with the .IC statements are all important techniques for transient simulation of oscillator circuits.

Interpolated vs. Noninterpolated Output Results

In Berkeley SPICE2G.6, the transient simulation output is interpolated from the actual timepoints SPICE computes and the print interval found on the transient statement (Fig. 5.1b). Berkeley SPICE2G.6 does not give users access to the actual timepoints where SPICE solves the circuit equations. Because of this, SPICE users must either set the print interval to a small value (which results in large output files) or set the print interval to a large value and risk the chance of missing parts of the transient response.

In most of the vendor-offered, SPICE-like simulators, users are given the choice of printing either the actual (noninterpolated) timepoints where SPICE solves the circuit or the interpolated output results. *It is highly recommended that SPICE users view the noninterpolated transient results whenever possible.* Simulate the disk circuit ch5-14a.cir with the command:

    SIM CH5-14A.CIR

Your results should match Fig. 5.14*a*. Now simulate the disk circuit ch5-14b.cir with the command:

```
SIM CH5-14A.CIR
```

Your results should match Fig. 5.14*b*.

The disk file ch5-14b.cir contains a listing identical to the ch5-14a.cir circuit with the addition of the *RSPICE specific* option NOINTR (.OPTION NOINTR). The NOINTR (NO INTeRpolate) option suppresses the interpolation routine and prints the actual timepoints at which RSPICE solves the circuit equations. Users should note that the NOINTR is not a standard SPICE option and each of the SPICE-like simulators may use a different option or command to suppress the interpolation routine.

Timestep Control in Other Simulators

Hspice

Hspice uses a proprietary timestep control algorithm named DVDT. This algorithm senses the rate of change of voltage levels in the circuit and adjusts the timestep accordingly. Hspice also offers the standard Local Truncation Error and Iteration-Count timestep control algorithms.

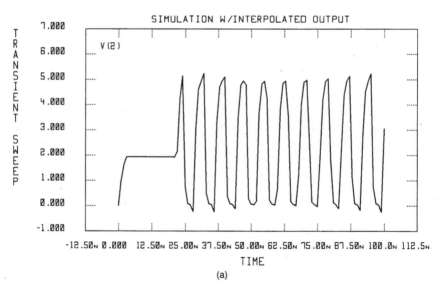

(a)

Figure 5.14(a)   (*Reprinted from* Successfully Simulating Circuits With SPICE. *Used with permission.*)

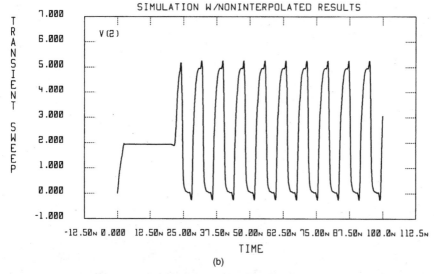

Figure 5.14(b)   Interpolation often hides the finer details of a simulation. Use the actual solution points (noninterpolated data) whenever possible. (*Reprinted from* Successfully Simulating Circuits With SPICE. *Used with permission.*)

All three algorithms benefit from the following enhancements. First, Hspice automatically limits the maximum timestep to RMAX times the print interval. RMAX is a user-selectable option which has a default value of 2. Second, an improved breakpoint adjustment decreases the chance that Hspice will skip a source breakpoint.

In Hspice, the POST option forces the simulator to store the noninterpolated results in a raw data file for later viewing. If the option INTRPL is set on the option statement, Hspice saves the interpolated results in the raw data file. The transient results printed in the output file are the interpolated values of the output.

IS_Spice

IS_Spice uses the standard SPICE2G.6 timestep control algorithms. Both the Iteration-Count and Local Truncation Error methods are available. The Local Truncation Error algorithm is the default method in IS_Spice.

IS_Spice does not offer a mechanism to generate the noninterpolated output results during transient analysis.

Micro-Cap IV

Micro-Cap IV uses a modification of the Local Truncation Error timestep control algorithm. This algorithm is fundamentally identical to

SPICE's algorithm, with the addition of an enhancement which improves the algorithm's ability to detect source breakpoints. This improvement helps ensure that Micro-Cap IV will not skip source breakpoints.

Because of this improvement, Micro-Cap IV does not offer the Iteration-Count timestep control algorithm as a user-selectable option.

In Micro-Cap IV, the simulator saves and graphs the noninterpolated transient results. Micro-Cap IV does not generate a standard SPICE output file.

Pspice

Pspice uses a modification of the Local Truncation Error timestep control algorithm. This algorithm is fundamentally identical to SPICE's algorithm, with the addition of two enhancements which improve the algorithm's performance. The first improvement is an automatic limit to the maximum timestep whenever a sine wave source is in use. In Pspice, the maximum timestep is automatically limited to $1/(8*FREQ)$, where $FREQ$ is the highest frequency of any sinusoidal generator in the circuit. The second improvement is a change in the breakpoint calculation, which ensures that Pspice will not skip source breakpoints.

Because of these improvements, Pspice does not offer the Iteration-Count timestep control algorithm as a user-selectable option.

In Pspice, the .PROBE statement forces the simulator to store the noninterpolated transient results in a raw data file for later viewing. The transient results printed in the output file are the interpolated results of standard SPICE.

Summary

The dynamic timestep control algorithms greatly improve the speed and accuracy of the transient simulation. But the timestep control algorithms are not foolproof, especially the (default) Local Truncation Error algorithm. Too often, the timestep becomes too large for a given circuit response. The resulting simulation often contains a large percentage of error. Sometimes, the timestep will become large enough to completely skip important behaviors of the circuit.

To guarantee high-quality simulation results, SPICE users need to know which timestep control algorithm to use, how to manually limit the maximum timestep with the HMAX parameter on the transient statement, and how to view the noninterpolated transient output results when available.

References

1. L. W. Nagel, *SPICE2: A Computer Program to Simulate Semiconductor Circuits,* Electronics Research Laboratory Rep. No. ERL-M520, University of California, Berkeley, 1975.
2. *SPICE Version 2G.6 User's Guide.*

SPICE Options

One of the most important commands in the SPICE input file is the .OPTIONS statement. The .OPTIONS statement is used to set parameter values which control the simulation. The parameters alter the numeric algorithms of SPICE, change the way the simulation results appear in the output file, generate additional information about the simulation or the netlist, as well as perform a variety of miscellaneous functions.

Understanding the .OPTIONS statement parameters is crucial to producing fast, accurate, convergent SPICE simulations. While the default parameter settings produce accurate results on some circuits, many others will require adjustments to one or more parameters. Remember SPICE was originally optimized for integrated circuit-level voltages and currents. Board- and system-level designers and power supply designers have already seen several parameter values which must be changed simply because of the magnitude of voltages and currents flowing through their circuits. Circuits which exhibit nonconvergent behavior can be forced to a convergent state by altering the parameters which control the numeric algorithms of SPICE. Transient simulations which display timestep control anomalies or numeric integration instabilities can be corrected by applying the proper parameter settings. Chapters 3, 4, and 5 all focused on producing faster, more accurate, convergent simulations. In these chapters, specific problems were corrected with the application of one or more .OPTIONS parameters. The .OPTIONS statement serves as an avenue to both correcting circuit simulation problems and optimizing simulation run time. For fast, accurate, convergent simulations, understanding the .OPTIONS statement is not just desirable, it is a requirement.

SPICE has 34 user-selectable options. Most of the vendor-offered SPICE-based simulators use the same options and often add many new ones. Of the 34 available, 33 are documented in the Berkeley SPICE2G.6 User's Guide.[1] One was never added to the original documentation. Of these 34, 6 operate as flags and signal an event when the option is named on the .OPTIONS line. As an example, to print accounting information (run time, number of iterations used, etc.) for your simulation run, a user would specify .OPTIONS ACCT. The other 29 operate as parameters and require that a numeric value be assigned on the .OPTIONS line. For example, to set the number of iterations allowed during the DC bias-point calculation, a user would specify .OPTIONS ITL1=X, where X is the number of iterations allowed.

The proper command syntax for the .OPTIONS statement is:

```
.OPTIONS <option1 option2 option3 ...>
```

However, because of the way SPICE processes the lines of the input file, only the first three characters in the .OPTIONS command are required. All of the following are acceptable forms of the .OPTIONS statement.

```
.OPTIONS <option1 option2 option3 ...>
.OPTION <option1 option2 option3 ...>
.OPTIO <option1 option2 option3 ...>
.OPTI <option1 option2 option3 ...>
.OPT <option1 option2 option3 ...>
```

Users may specify more than one .OPTIONS line in an input file. If the same option is declared more than once, only the last value of the option will be used. This is important in simulations which are composed of different blocks or subsections, each of which may contain different option statements. SPICE does not allow different sections of a netlist to operate with specific option settings. Once set, an option applies globally to the circuit until another .OPTIONS statement resets the value.

All of the options have names composed of two to six alphanumeric characters. Some of the names were well chosen and clearly express the meaning of the option. For example, the option which selects the numeric integration method used in the simulation, METHOD= 'GEAR', is easy to understand and use. But often the names are less than self-explanatory. For example, the option which selects which timestep control algorithm is used in the simulation is LVLTIM=X. Lastly, there are just a few options which have names which appear to be explicit and meaningful but do not operate as the name implies. For example, RELTOL, ABSTOL, and VNTOL all determine the accuracy of the simulation results, as might be implied from the TOLerance portion of the name. The options CHGTOL and TRTOL are both correction factors for a shortcoming in the Local Truncation Error timestep control algorithm. They do not have a direct effect on the accuracy of the simulation results even though they both have TOL in their names. A clear understanding of each of the options helps build a strong foundation for high-quality simulation.

As a fledgling SPICE user, the author recalls trying to decode the mysterious one-line description of each option in the SPICE User's Guide.[1] Since this young engineer had little idea of what many of the options did, many misbehaving simulations were often liberally doused with a variety of different options. With each simulation, the young engineer anxiously awaited the results, each time hoping that these option settings would be the key to success. Often, hours of simulation time were consumed before stumbling on the right combination. Too many times, the right combination was never found.

SPICE Option Definitions

In this section, each of the 34 standard SPICE options will be defined in clear detail. Where applicable, formulas showing how each of the option values should be set for specific circuits will be given. Some options should always be set regardless of the type of simulation. Others may only be needed occasionally. With each option, recommendations will be made as to whether the option should always be set and/or when it should be used.

For designers using one of the SPICE-like simulators, most of the options detailed here have a one-to-one correspondence with the same named option in your simulator. Appendix A lists the .OPTIONS parameters for several popular circuit simulators and, where possible, shows the correspondence between the SPICE2G.6 option and the vendor-offered option.

The first 8 options may be classified as options which control the output file format, the next 18 control the numeric algorithms of SPICE, 4 determine the default MOS gate geometries, and finally, 4 more can only be classified as miscellaneous options. The options will be presented in this order.

Output File Format Options

The output file format options include ACCT, LIST, NOMOD, NOPAGE, NODE, OPTS, LIMPTS=X, and NUMDGT=X.

.OPTIONS ACCT

The accounting option is a flag and directs SPICE to print 47 run-time statistics about the job just simulated. The results appear in the output file. The accounting section of the output file contains an abundance of information about the simulation run. Within the accounting section, users will find information relating to the speed and efficiency of the simulation, the time and number of iterations each analysis type consumes, the size and number of elements in the circuit, the size of the solution matrix, the number of analysis points, and the maximum number of nodes or circuit elements which may be simulated.

Figure 6.1 shows an example of the accounting information as it appears in the output file. The information is segmented into two different forms, numeric values pertaining to elements of the simulation and run-time statistics.

The first row of values in the accounting section relates to the elements in the circuit. The second row displays information regarding the analysis. The third reveals details of the system equations, and the final row contains data about the transient analysis and the memory requirements of the simulation.

The headings which describe the circuit element, analysis, system equation, and transient and memory entries are shown in Tables 6.1a, b, c, and d. The items marked in italics will be the most useful to simulation users.

Following the four rows of data are four columns of data displaying the run-time statistics and number of iterations each program segment used. The program segment headings are shown in Table 6.2. Again, the headings shown in italics will be of interest to most users.

Two of the best uses for the information found in the accounting section are to improve the speed or efficiency of a simulation and to determine the maximum circuit size or maximum analysis time supported by your simulator.

Using the accounting information to improve simulation. To improve the speed of simulation, observe the number of iterations each of the major analysis types consumes and use this as a gauge as you alter one or more options or commands of the simulation. For example, Table 6.3 shows the number of iterations required and the amount of time required to compute the dc bias point of disk circuit ch6-3.cir. For this circuit, adding the .NODESET statement reduced the number of iterations from 53 to 13 and reduced the time used for the bias point analysis by 56 percent. Try adding one or more .NODESETs to your circuit and watch the number of iterations listed under the DCAN program segment. With the proper settings, you can often reduce the number of

```
PWM SIMULATION

****   JOB STATISTICS SUMMARY          TEMPERATURE =   27.000 DEG C

***********************************************************************

NUNODS NCNODS NUMNOD NUMEL DIODES  BJTS  JFETS  MFETS

   3     8     10    15     2      0     0      0

NUMTEM ICVFLG JTRFLG JACFLG INOISE  IDIST   NOGO

   1     0    101     0      0      0      1

NSTOP   NTTBR   NTTAR   IFILL   IOPS   PERSPA

 16.     36.     38.     2.     59.    85.156

NUMTTP  NUMRTP  NUMNIT  MAXMEM  MEMUSE  COPYKNT

 708.    94.    2920.  1000000   9408    51948.


         PROGRAM SEGMENT        TIME    NUMB. OF ITERATIONS

         READIN              0.49

         SETUP               0.17

         TRCURV              0.00          0.

         DCAN                0.22         30.

         DCDCMP      2.640                 2.

         DCSOL       1.980

         ACAN                0.00          0.

         TRANAN             17.08       2920.

         OUTPUT              1.60

         LOAD        6.210

         CODGEN      0.000                 0.

         CODEXC      0.000

         MACINS      0.000

         OVERHEAD            0.16

         TOTAL JOB TIME     19.72
```

Figure 6.1   ACCT option output.

TABLE 6.1(*a*)   Circuit Element Information

NUNODS	This is the number of nodes in the main circuit.
NCNODS	This is the number of circuit nodes after subcircuit expansion.
NUMNOD	This is the number of nodes after subcircuit expansion and includes the internal parasitic nodes for semiconductor devices.
NUMEL	This is the number of elements (components) in the circuit.
DIODES	This is the number of diodes in the circuit.
BJTS	This is the number of BJTs in the circuit.
JFETS	This is the number of JFETs in the circuit.
MFETS	This is the number of MOSFETs in the circuit.

TABLE 6.1(*b*)   Analysis Information

NUMTEM	This is the number of temperatures at which the circuit will be analyzed.
ICVFLG	This is the number of output points in a DC sweep analysis.
JTRFLG	This is the number of output points in a transient sweep analysis.
JACFLG	This is the number of output points in an AC frequency sweep.
INOISE	This flag indicates whether a noise analysis was performed.
IDIST	This flag indicates whether a distortion analysis was performed.
NOGO	This flag indicates whether an error condition was encountered during simulation. A number of different error conditions may set this flag.

TABLE 6.1(*c*)   Circuit Element Information

NSTOP	This is the number of nodes in the main circuit matrix $(N \times N)$.
NTTBR	This is the number of matrix entries including subcircuit expansion and internal nodes.
NTTAR	This is the number of matrix entries including subcircuit expansion and internal nodes after "fill-in" is complete.
IFILL	This is the number of "fill-ins" generated in the conductance matrix.
IOPS	This is the number of floating-point operations required to solve the system equations.
PERSPA	This is the percent of the conductance array which was empty. Electronic circuits generally lead to extremely sparse system equations. While PERSPA may not be useful directly, it is interesting to observe this number for a variety of circuit types.

TABLE 6.1(*d*)   Transient Analysis and Memory Information

NUMTTP	This is the number of internal timepoints SPICE used during the transient analysis. This is not the same as JTRFLG, which is the number of interpolated output points.
NUMRTP	This is the number of times SPICE failed to converge, cut the timestep, and reattempted the circuit solution during transient analysis. For an efficient run, NUMRTP should be less than 10 percent of NUMTTP. In most simulations, NUMRTP may be reduced (transient efficiency increased) by raising ITL4.
NUMNIT	This is the total number of iterations used during the transient simulation.
MAXMEM	This is the maximum amount of memory available to SPICE. The units of memory are 4-byte words.
MEMUSE	This is the amount of memory used during the simulation. The units of memory are 4-byte words. Comparing this number to MAXMEM enables users to determine the maximum circuit size or the maximum run time a particular version of SPICE can support.
COPYKNT	This is the number of times the copy routine was called in SPICE. The copy routine is used to move numeric values throughout the program.

TABLE 6.2   Program Segment Headings

READIN	This segment reads the input file listing, parses the input file syntax, and stores the circuit elements in memory.
SETUP	This segment defines the system equations and initializes the circuit sources.
TRCURV	This segment performs the DC sweep analysis.
DCAN	This segment performs the DC operating point (bias point) calculation.
DCDCMP	This segment performs the numerical pivoting and LU factorization of the system equations.
DCSOL	This segment performs the back substitution of the system equations.
ACAN	This segment performs the AC small-signal frequency sweep analysis.
TRANAN	This segment performs the transient time sweep analysis.
OUTPUT	This segment prints the simulation results to the output file.
LOAD	This segment loads the device conductance and current values into the solution array during the iterative solution procedure.
CODGEN	This segment generates machine code to solve the solution array. This routine is called with the LVLCOD option. Most versions of SPICE have disabled this option.
CODEXC	This segment solves the solution array when the LVLCOD option is selected. Most versions of SPICE have disabled this option.
MACINS	This segment builds the machine instructions to solve the solution array when the LVLCOD option is selected. Most versions of SPICE have disabled this option.
OVERHEAD	This is time consumed by SPICE for general housekeeping chores.

TABLE 6.3   Bias Point Run-Time Statistics

	Number of iterations	Time
Without NODESET	53	4.29 sec.
With NODESET	17	1.87 sec.

iterations by 50 percent or more and eliminate a considerable portion of the time required for the operating point calculation.

For transient analysis, watch the values of NUMTTP, NUMRTP, and NUMNIT. NUMTTP is the number of timepoints SPICE uses during the transient simulation; this is not the number of points printed in the output file. NUMRTP is the number of timepoint reversals or the number of times SPICE cuts the timestep and steps back in time to reattempt the solution point after failing to converge on a given timepoint. NUMNIT is the total number of iterations required to complete the transient analysis. By raising the value of ITL4 on the .OPTIONS statement, SPICE will cut the timestep less often during transient analysis. The less timepoint reversals SPICE uses, the faster the simulation will complete. For an efficient simulation, the values of NUMTTP, NUMRTP, and NUMNIT should be as low as possible. In general, the values of NUMTTP, NUMRTP, and NUMNIT may be decreased by raising ITL4. As you raise ITL4 in your simulation, watch the values of NUMTTP, NUMRTP, NUMNIT, and the amount of time used for the transient simulation. Continue raising ITL4 until no further improvement in the NUMTTP, NUMRTP, and NUMNIT can be achieved, the simulation time ceases to be reduced, or accuracy is degraded.

The disk file ch6-4.cir simulates the transient response of an SR flip-flop. In the ch6-4.cir file, the original circuit description has been duplicated several times with different values of ITL4. Table 6.4 shows the ITL4 value, the number of transient timepoints, the number of timepoint reversals, the number of transient iterations, and the overall simulation time. Notice that raising ITL4 beyond 40 produces no sig-

TABLE 6.4   Transient Run-Time Statistics

ITL4	NUMTTP	NUMRTP	NUMNIT	Job time
10 (default)	257	38	1012	17.47 sec.
15	136	7	561	9.77 sec.
20	112	1	471	8.29 sec.
40	108	0	417	7.53 sec.
100	108	0	417	7.53 sec.

Results obtained from a 33-MHz 386/387 computer.

nificant improvement in simulation speed. Simulate this circuit with RSPICE to verify these results.

To determine the maximum number of nodes or elements allowed in your SPICE program, watch the MAXMEM and MEMUSE values. SPICE uses a large data array to store the circuit description, the element values, and all of the model parameters. The same data array stores all the analysis results. The size of the array determines the maximum number of elements or the maximum number of analysis points allowed for a given program. If the storage requirements of your simulation ever exceed the amount of memory available, SPICE will print a most unfriendly message in the output file and abort the simulation. MAXMEM is the maximum amount of memory your version of SPICE can address. MEMUSE is the amount of memory used during a simulation. The data array must hold both the elements of a circuit and all of the analysis data. Because of this, SPICE can simulate the largest number of elements when performing a very short analysis, say a dc operating point calculation. Conversely, SPICE can simulate the longest analysis (most analysis points) when a very small circuit is being simulated.

To determine the maximum number of elements SPICE can simulate, create a circuit netlist with several elements. Include the ACCT option in the circuit netlist. Simulate the circuit, record the MAXMEM and MEMUSE values, and add several additional elements to the original netlist. Again simulate the circuit and record the new MEMUSE value. The difference between the two MEMUSE values represents the amount of memory required for the additional elements. The amount of memory per element and MAXMEM can be used to determine the theoretical maximum number of elements SPICE can simulate. The disk file named ch6-5.cir contains a 7-stage and a 14-stage ring oscillator. When simulated, the MEMUSE value can be used to calculate the amount of memory required for the seven additional stages of the circuit. Table 6.5 shows the amount of memory required to simulate each circuit. The addition of seven extra stages required 4376 memory units (or 625 memory units per inverter stage). By extrapolating the memory units per stage to the 200,000 memory units available, the user may deduce that up to approximately 320 stages (320 inverters) could be used in this analysis!

TABLE 6.5   Run Statistics for Maximum Element Calculation

Oscillator stages	MEMUSE	MAXMEM
7	4720	200,000
14	9096	200,000

Obviously, the maximum number of elements will be reduced when a more sophisticated analysis is performed.

A similar calculation can be made for the maximum simulation length of any job. For example, to determine if a long transient simulation may be performed, run a short transient simulation on the circuit. Then increase the transient duration by 20 percent over the previous run. The difference in MEMUSE values represents the amount of memory required for the 20 percent increase in transient duration. The maximum transient duration can be calculated from the MAXMEM value and the amount of memory used for the longer transient run.

The disk file named ch6-6.cir contains the same 7-stage ring oscillator of the previous problem. The first circuit file simulates the transient behavior for 4nS (with a 100pS print resolution). The second simulates the behavior for 8nS (again with a 100pS print resolution). Table 6.6 shows the amount of memory the two runs use.

From the amount of memory per nanosecond of transient duration, the user may deduce that the same 100pS print resolution for this circuit will yield a maximum transient duration of 18uS.

For the purist, both MAXMEM and MEMUSE have units of four-byte words; so, to determine the actual number of bytes of memory used for either of these, multiply the MAXMEM and MEMUSE values by 4. The result is the number of bytes SPICE is using.

.OPTIONS LIST

The LIST option is a flag and directs SPICE to print a list of all the element types of a simulation in the output. The list is ordered by element type. All resistors are listed first, then all capacitors, inductors, etc., until all the elements in the circuit have been listed. Included in the list are the nodal connections for each element, the element value, any optional element parameters, and subcircuit connectivity information. Figure 6.2 is an example of the LIST option.

The LIST option is best used in diagnosing topology-related errors. Erroneous node connections can quickly be uncovered and resolved. The list is ordered by element type, which makes node connections much easier to read.

TABLE 6.6   Run Statistics for Maximum Run-Time Calculation

Transient duration	MEMUSE	MAXMEM
4nS	4720	200,000
8nS	4764	200,000

THIS IS A SIMPLE X STAGE RING OSCILLATOR

****   CIRCUIT ELEMENT SUMMARY   TEMPERATURE =   27.000 DEG C

****

**** RESISTORS

NAME	NODES		VALUE	TC1	TC2
RUP.X1	1000	2	1.43E+03	0.00E+00	0.00E+00
RUP.X2	1000	3	1.43E+03	0.00E+00	0.00E+00
RUP.X3	1000	4	1.43E+03	0.00E+00	0.00E+00
RUP.X4	1000	5	1.43E+03	0.00E+00	0.00E+00
RUP.X5	1000	6	1.43E+03	0.00E+00	0.00E+00
RUP.X6	1000	7	1.43E+03	0.00E+00	0.00E+00
RUP.X7	1000	1	1.43E+03	0.00E+00	0.00E+00

**** CAPACITORS AND INDUCTORS

NAME	NODES		IN COND	VALUE
COUT.X1	2	0	0.00E+00	2.00E-14
COUT.X2	3	0	0.00E+00	2.00E-14
COUT.X3	4	0	0.00E+00	2.00E-14
COUT.X4	5	0	0.00E+00	2.00E-14
COUT.X5	6	0	0.00E+00	2.00E-14
COUT.X6	7	0	0.00E+00	2.00E-14
COUT.X7	1	0	0.00E+00	2.00E-14

**** INDEPENDENT SOURCES

NAME	NODES		DC VALUE	AC VALUE	AC PHASE	TRANSIENT
VDD	1000	0	5.00E+00	0.00E+00	0.00E+00	

**** MOSFETS

NAME	D	G	S	B	MODEL	W	AD	PD	RDS
						L	AS	PS	RSS
M1.X1	2	1	0	0	NCHAN	1.3E-04	0.0E+00	0.0E+00	1.0E+00
						3.0E-06	0.0E+00	0.0E+00	1.0E+00
M1.X2	3	2	0	0	NCHAN	1.3E-04	0.0E+00	0.0E+00	1.0E+00
						3.0E-06	0.0E+00	0.0E+00	1.0E+00
M1.X3	4	3	0	0	NCHAN	1.3E-04	0.0E+00	0.0E+00	1.0E+00
						3.0E-06	0.0E+00	0.0E+00	1.0E+00

Figure 6.2   LIST option output.

.OPTIONS NOMOD

The NOMOD option is just one of many options which were added to SPICE early in the development of the program to save computer resources. During the dark ages when a central computer center, batch jobs, and line-printer outputs were the norm, the authors of SPICE realized the output file contained many bits of information which were not always needed. One of these was the model parameter information. Whenever SPICE encounters a .MODEL statement in the output file, the program passes each model parameter into a parser. The parser determines whether the model parameter name and value are valid. Once done, SPICE prints the model parameter name and value in the output file. With each model parameter name and value on a double-spaced line, simulations with more than a few .MODEL statements resulted in page after page of model parameter values. To allow users to suppress the parser print-back of the model parameters and ultimately save output paper (remember everything was printed to the line printer in the early days), the authors of SPICE installed the NOMOD option.

NOMOD is a flag and, when set, the model parameters will not be reprinted during the model parameter parsing stage. Figure 6.3 shows an example of the model parameter parsing section which is suppressed with the NOMOD option.

.OPTIONS NOPAGE

NOPAGE is similar to the NOMOD option in that the authors of SPICE installed NOPAGE to reduce or eliminate redundant information in the output file. NOPAGE is a flag which, when set, suppresses the page-feeds and much of the header information in the output file.

When SPICE performs a simulation, a 10-line header is printed at the start of each new section in the output file. The header contains information about the date, time, temperature, section title, and simulation title. The header also contains a page-feed at the beginning of each page so that, when printed, the header aligns with the top of the page. Often, this leaves much of the previous page blank. The NOPAGE option suppresses the page-feeds and reduces the 10-line header to a single-line header which describes the section title.

Today, most simulation outputs are stored in files rather than being directed to the lineprinter, and, because file space (disk space) is relatively cheap and abundant, the NOMOD and NOPAGE options play a greatly reduced role in simulation.

.OPTIONS NODE

The NODE option is a flag that is similar to the LIST option because it directs SPICE to print netlist information in the output file. But where

```
THIS IS A SIMPLE X STAGE RING OSCILLATOR


****    MOSFET MODEL PARAMETERS            TEMPERATURE =   27.000 DEG C


********************************************************************


                NCHAN

TYPE            NMOS

LEVEL           3.000

VTO             0.864

KP              4.74E-05

GAMMA           0.306

PHI             0.588

LAMBDA          0.00E+00

RD              1.28E+01

RS              1.28E+01

PB              0.800

CGSO            3.93E-10

CGDO            3.93E-10

CGBO            5.00E-10

CJ              1.09E-04

MJ              0.333

CJSW            1.60E-10

MJSW            0.333

TOX             5.00E-08

NSUB            4.27E+15

TPG             1.000

XJ              8.69E-07

LD              4.14E-07

UO              511.070

UCRIT           0.00E+00

UEXP            0.000
```

Figure 6.3   Model parameter output.

LIST generates a list of components in the circuit netlist, NODE gen-erates a list of the nodes in the circuit and the elements connected to each node. Figure 6.4 shows an example of the node connection infor-mation printed in the output file.

Although the NODE option can be used to debug topology-related problems, NODE does not describe which terminal of a given compo-nent is connected to a specific node. For this reason, the LIST option is a much better topology diagnostic tool. A better use of the NODE option is assisting in the resolution of a dc operating point nonconver-gence. If a circuit fails to converge during a dc operating point analysis, SPICE prints a list of the last iterative node voltage values in the out-put. Often, one or more of these voltage values are extremely, unrealis-tically high and obviously wrong. When one or more of these nodes are encountered, use the NODE option to generate a list of all the elements connected to that node. One or more of those elements is generating the nonconvergence problem. Once the problem elements have been identified, use the .NODESET or OFF statements presented in Chap. 3 to correct the nonconvergence problem.

```
THIS IS A SIMPLE X STAGE RING OSCILLATOR


****    ELEMENT NODE TABLE              TEMPERATURE =   27.000 DEG C


************************************************************************


    0    COUT.X1  COUT.X2  COUT.X3  COUT.X4  COUT.X5  COUT.X6  COUT.X7
         VDD      M1.X1    M1.X1    M1.X2    M1.X2    M1.X3    M1.X3
         M1.X4    M1.X4    M1.X5    M1.X5    M1.X6    M1.X6    M1.X7
         M1.X7

    1    RUP.X7   COUT.X7  M1.X1    M1.X7

    2    RUP.X1   COUT.X1  M1.X1    M1.X2

    3    RUP.X2   COUT.X2  M1.X2    M1.X3

    4    RUP.X3   COUT.X3  M1.X3    M1.X4

    5    RUP.X4   COUT.X4  M1.X4    M1.X5

    6    RUP.X5   COUT.X5  M1.X5    M1.X6

    7    RUP.X6   COUT.X6  M1.X6    M1.X7

 1000    RUP.X1   RUP.X2   RUP.X3   RUP.X4   RUP.X5   RUP.X6   RUP.X7
         VDD
```

Figure 6.4   NODE option output.

.OPTIONS OPTS

The OPTS option is one of the most useful options in SPICE. The OPTS option is a flag that directs SPICE to print the value of each of the optional parameters during a SPICE simulation. At first, this seems quite redundant, because users must specify options on the .OPTIONS line. But the OPTS option prints all user-selectable options, not just those specified on the .OPTIONS statement.

One of the best uses of the OPTS option is to determine the default setting of all the SPICE options. Determining the default settings of any SPICE simulator is extremely important because different simulators will assign different default values that will have a dramatic impact on the speed, accuracy, and convergence properties of the simulator.

To determine the default settings in SPICE, generate a simple netlist with a .OPTIONS line which has nothing but the OPTS option declared. After simulation, the output file will contain a section similar to the list shown in Fig. 6.5. These are the default option settings of SPICE.

If you use the OPTS during a simulation with one or more option parameters set on the .OPTIONS line (in addition to the OPTS statement), the declared value, rather than the default value, will be printed in the output file.

.OPTIONS LIMPTS=X

The LIMPTS=X is the parameter that sets an upper limit on the number of solution points which can be simulated or printed in the output file. LIMPTS is one of several safety nets which were installed in SPICE to prevent users from accidentally consuming excessive computer resources. For example, if the following line

```
.TRAN .1MS 20MS
```

was accidentally entered as

```
.TRAN .1NS 20MS
```

SPICE would try to simulate 200 million data points!

Excessive run times or output files are not a significant problem for today's single-user computers. Workstation or personal computer users can easily abort a long simulation run without significant repercussions. But when SPICE was written, a central computing facility, batch jobs, and line printer outputs were used. Often, users knew nothing about how their job was running until the printed output was carried

```
****** 11/24/90 ******* RSPICE V3.2.0 04/15/92 ******  6:13 pm ********


OPTIONS EXAMPLE


****    OPTION SUMMARY                 TEMPERATURE =   27.000 DEG C


************************************************************************


DC ANALYSIS -

    GMIN   =  1.000E-12
    RELTOL =  1.000E-03
    ABSTOL =  1.000E-12
    VNTOL  =  1.000E-06
    LVLCOD =      1
    ITL1   =    100
    ITL2   =     50

    PIVTOL = .1.000E-13
    PIVREL =  1.000E-03

TRANSIENT ANALYSIS -

    METHOD =  TRAP
    MAXORD =      2
    CHGTOL =  1.000E-14
    TRTOL  =  7.000E+00
    LVLTIM =      2
    MU     =    0.500
    ITL3   =      4
    ITL4   =     10
    ITL5   =   5000


MISCELLANEOUS -

    LIMPTS =    201
    LIMTIM =      2
    CPTIME = 100000000
    NUMDGT =      4
    TNOM   =   27.000
    DEFL   =  1.000E-04
    DEFW   =  1.000E-04
    DEFAD  =  0.000E+00
    DEFAS  =  0.000E+00
```

Figure 6.5   OPTS option output.

to the output room. The second .TRAN statement generates 200 million output points in the output file. Assuming 66 lines per page and 100 pages in a one-inch stack of fanfold printer paper, 200 million points translates into a stack of printer paper just short of a half mile high! Just imagine the look on the computer center person that tried to take that stack off the printer!

By default, LIMPTS is set to 201 in SPICE. Many vendor-offered versions of SPICE reset this to a higher value. During the reading of the input file, SPICE determines the number of points the user requested for the simulation. If the number of points is greater than LIMPTS, SPICE prints an error message in the output file and terminates without performing the simulation. The user can reset the parameter to either the exact number of points he or she wishes to simulate or simply to a larger default value. (The author usually sets LIMPTS to 10,000 in most simulations.)

.OPTIONS NUMDGT=X

The last of the output options is the NUMDGT=X parameter. NUMDGT is the number of significant digits for voltages and currents printed in the output file. Changing this option alters the print format used by SPICE, not the accuracy of the simulator. By default, NUMDGT is set to 4. Allowable values for this parameter are 1 to 7.

Output File Format Options Summary

The output file format options determine the type and form of information printed in the output file. The information obtained may be used to debug netlist-related problems, minimize output file size, determine the default SPICE option settings, and determine the maximum circuit size or run time.

Numeric Control Options

The numeric control options consist of four subgroups. In the first group are the convergence parameters GMIN, RELTOL, ABSTOL, and VNTOL. The second group contains the iteration limits ITL1, ITL2, ITL3, ITL4, ITL5, and ITL6. The transient analysis options group follows with LVLTIM, METHOD, MAXORD, MU, TRTOL, and CHGTOL. Finally, two options define the matrix pivoting group, PIVTOL and PIVREL. All of the numeric control options are parameterized options rather than flags. *Parameterized* simply means that the value of the option must follow the option name.

Convergence options

The convergence options include GMIN, RELTOL, ABSTOL, and VNTOL. Chapter 3 describes each of these and discusses how they impact the accuracy and convergence characteristics of a circuit.

.OPTIONS GMIN=X. The parameter GMIN=X represents the minimum allowed conductance (maximum resistance) of any element in the circuit. GMIN is also the value of small conductance (large resistance) placed in parallel with every semiconductor pn junction in the circuit. GMIN can be equated with the parasitic leakages associated with a given circuit. The default value of GMIN = 1e − 12 mhos.

The best use for GMIN is in aiding the convergence properties of the simulator. For many circuits, raising GMIN from the default will result in better overall convergence characteristics without sacrificing the accuracy of the result. To select a value for GMIN, determine the smallest parasitic resistance which could be placed across any two nodes of a circuit without affecting the normal circuit operation. Set GMIN=X to the inverse of the resistance value.

$$\text{GMIN} = \frac{1}{(\text{smallest parasitic resistance})}$$

.OPTIONS RELTOL=X. The RELTOL=X parameter is the relative error tolerance required for convergence. To solve the circuit equations, SPICE starts the Newton-Raphson iterative solution algorithm. The Newton-Raphson algorithm iterates over and over, searching for a set of node voltages and branch currents which will make the remaining circuit equations conform to Kirchhoff's voltage and current laws. Since the Newton-Raphson algorithm never really "knows" the proper circuit voltages and currents, the signal to stop iterating must come from outside the algorithm. One of the useful properties of the Newton-Raphson algorithm is that when the algorithm is close to the exact solution, the change in iteration-to-iteration voltage and current values approaches zero.

Because digital computers may have a difficult time predicting when the difference in voltage and current is exactly zero (because of round-off error), SPICE presumes the Newton-Raphson algorithm has found the exact solution when the difference between iterative voltages and currents is less than a given error tolerance. The first part of the error tolerance is a percentage change between iterative voltages and current. SPICE stops the iterative procedure when the percentage change between iterative voltage values for each node in the circuit is less than RELTOL and when the percentage change between iterative semiconductor branch currents is less than RELTOL.

RELTOL=X sets the relative error tolerance allowed in SPICE. The default value of RELTOL = .001, or one-tenth of one percent. For most circuits this is a good compromise between speed and accuracy. More accuracy requires more iterations at a given solution point. More iterations translate into more simulation time. Less accuracy requires less iterations, resulting in faster simulation execution. Nagel demonstrated that for each additional significant digit of accuracy required, the number of iterations required to achieve the solution doubles.[2] For most analog circuits, the default value of RELTOL yields satisfactory simulation accuracy and speed. For many digital circuits, RELTOL may be raised to .01 or .05 for faster simulation execution without a significant change in accuracy. Experiment with RELTOL for different circuits. Use a value which results in the best compromise between speed and accuracy. The value of RELTOL must be determined before the VNTOL=X and ABSTOL=X parameters may be set.

.OPTIONS VNTOL=X ABSTOL=X. The VNTOL=X and ABSTOL=X parameters work with the RELTOL option to determine the error tolerance for the iterative solution algorithm in SPICE. The percentage change criterion works well for most voltage and current values used in circuit simulation, except one. When the node voltage or branch current approaches zero, the percentage change error tolerance criterion fails to predict a good termination point for the iterative algorithm. Because of this, in addition to the percentage change criterion, a lower absolute error tolerance must also be determined. The absolute error tolerances will define convergence when the node voltage or branch current falls to zero. For node voltages in the circuit, VNTOL=X is the absolute error tolerance between iterations, and ABSTOL=X is the absolute error tolerance for semiconductor branch currents. These two option parameters determine the lower resolution on voltage and current for the simulator.

The default value for VNTOL is 1uV. The default value for ABSTOL is 1pA. The authors of SPICE set these to integrated circuit level voltages and currents. For any circuit which is not an integrated circuit (i.e., board level, discrete, and especially power circuits), the VNTOL and ABSTOL parameters should be reset to align with the voltage and current levels seen in the circuit.

To select values for VNTOL and ABSTOL with respect to a given circuit, study the circuit. Determine the lowest voltage magnitude in the circuit and multiply RELTOL by this value. The result is the setting for VNTOL. For ABSTOL, determine the lowest current magnitude in the circuit and multiply RELTOL by this value. The result is the setting for ABSTOL. These two equations align the relative and absolute error tolerances for SPICE. Users should notice that the values for VNTOL and ABSTOL are dependent on the value of RELTOL. Any modification to RELTOL should be followed with an alteration to ABSTOL and VNTOL.

Readers are reminded that all of the convergence parameters are discussed in detail in Chap. 3.

Iteration limit options

The next set of options are the iteration limits. Six parameters define the iteration limits for different analysis types. The iteration limits all begin with the three characters, ITL.

.OPTIONS ITL1=X. The ITL1=X parameter defines the upper limit on the number of iterations allowed to compute the dc operating point. The default value of ITL1 is 100. Empirically it can be shown that about 60 percent of all circuits converge within 100 iterations, 75 percent converge within 200 iterations, and 92 percent converge within 500 iterations. If more than 500 iterations are required for convergence, something in the circuit has probably been incorrectly connected or one or more nodes needs to be initialized with the .NODESET or .IC statements.

The best use for ITL1 is in improving the dc operating point convergence characteristics. For all circuits, start the simulation with ITL1 set to 500. Since ITL1 is the upper limit on the allowed number of iterations, simulations which do not require excessive iterations to determine the dc operating point are not affected by raising ITL1.

The number of iterations SPICE uses in computing the dc operating point, dc sweep, or transient analysis can be observed by using the ACCT option discussed earlier in this chapter.

.OPTIONS ITL2=X. The ITL2=X parameter defines the upper limit on the number of iterations allowed at each step in a dc sweep analysis. The default value of ITL2 is 50.

The best use for ITL2 is in aiding convergence during a dc sweep analysis. For circuits with high-gain switch points, such as flip-flops, comparators, op-amps, and triggers, set ITL2 to 200.

.OPTIONS ITL3=X. The ITL3=X parameter sets the lower limit on the number of iterations for the Iteration-Count timestep control algorithm. When the Iteration-Count method is being used in a transient simulation, SPICE monitors the number of iterations at each timepoint in the simulation. If the number of iterations is ever less than or equal to ITL3, the timestep control algorithm automatically doubles the step size before the next timepoint is calculated.

The default value for ITL3 is 4 iterations and was determined empirically.[2] Neither Nagel nor the author found any significant improvement for the Iteration-Count timestep control algorithm by either raising or lowering this value. For most circuits, ITL3 should remain at the default value.

.OPTIONS ITL4=X. The ITL4=X parameter sets the upper limit on iterations for either the Iteration-Count timestep control algorithm or the Local Truncation Error timestep control algorithm. If the number of iterations at a given timepoint is greater than ITL4, SPICE discards the current timepoint, cuts the timestep by a factor of 8, then reattempts the solution at the new timepoint. The default value of ITL4 is 10 iterations.

The best use for ITL4 is enhancing the convergence characteristics and increasing the speed of a transient simulation. ITL4 has a dramatic effect on transient simulations. By raising ITL4, you raise the number of iterations allowed before SPICE discards the current timepoint. Because of the asymmetric manner in which SPICE changes the size of the timestep, reducing the number of times the timestep is cut drastically increases simulation speed. The section on transient nonconvergence in Chap. 3 explains more about why raising ITL4 will improve both the speed and the convergence characteristics of a transient simulation.

For transient simulation, set ITL4 to 40. Use the values of NUMTTP, NUMRTP, and NUMNIT (see the ACCT option section for a discussion of NUMTTP, NUMRTP, and NUMNIT) and adjust ITL4 accordingly for improved speed and efficiency in transient simulation.

.OPTIONS ITL5=X. The ITL5=X parameter is another safety net placed in SPICE to limit the total number of iterations used for transient simulations. At each timepoint in a transient run, SPICE counts the number of iterations required for solution at that timepoint. If the total number ever exceeds ITL5, SPICE will stop the transient simulation, print the result up to that point in the simulation, then print a message in the output file exclaiming the need to raise ITL5 to continue a longer simulation. The default value of ITL5 is 5000 iterations.

Depending on the circuit, 5000 iterations may require only a few minutes of computer time. Many simulations require more than 5000 iterations during the transient simulation. Users have the choice of raising ITL5 to the approximate number of iterations required; simply setting ITL5 to 0 suppresses the limit on the total number of allowed iterations.

.OPTIONS ITL6=X. The parameter ITL6=X is both a flag and a parameter value. The default value of ITL6 is 0. If ITL6 is set to a nonzero value, SPICE discards the normal dc operating point calculation and replaces it with the source-stepping algorithm. The value declared for

ITL6 serves as the iteration limit for each step in the source-stepping algorithm. Many people wrongly believed ITL6 was related to the size or number of steps in the process; not so. The size of the step is fixed and cannot be changed. ITL6 only determines the number of iterations allowed at each step.

The best use for ITL6 is in assisting in the dc operating point solution for circuits where the .NODESET statement cannot be used. Using .NODESETs is the preferred way to find the dc operating point, but .NODESETs cannot be applied to nodes within a subcircuit definition. For many large circuits, bistable nodes (such as flip-flops) cannot be accessed with the .NODESET statement. In these cases, raising ITL6 to 400 often helps achieve convergence.

Using the iteration limits wisely is one of the pieces of the puzzle to producing fast, accurate, convergent simulations. Another part of the puzzle is the transient analysis options. These are covered in the next section.

Transient analysis options

For most designers, transient analysis is the most often used analysis type in SPICE. Transient analysis is also the most complicated analysis mode and therefore the most error-prone type of analysis in SPICE. To produce accurate transient simulation, the user must understand the limitations of the transient analysis algorithms including numeric integration, timestep control, and the six SPICE options which are applicable only to transient analysis.

.OPTIONS LVLTIM=X. The parameter LVLTIM=X defines which timestep control algorithm SPICE uses during the transient simulation. By default LVLTIM is set to 2, which calls the Local Truncation Error (LTE) timestep control algorithm. Although slightly faster, the LTE timestep control algorithm is more error-prone than the Iteration-Count method. Without applying a limit to the maximum internal timestep, the LTE method may yield poor results when simulating asynchronous, sinusoidal, or inductive circuits. For these circuits, setting LVLTIM to 1 selects the Iteration-Count (IC) timestep control algorithm. The IC timestep control algorithm automatically limits the size of the internal timestep and may be much more reliable on these types of circuits. For a more detailed discussion of LVLTIM=X and timestep control algorithms, see Chap. 5 in this text.

.OPTIONS METHOD='yyyy'. The METHOD='YYYY' option defines the numeric integration method used to calculate capacitor currents and inductor voltages. Valid settings are METHOD='GEAR' or METHOD=

'TRAP'. The default is TRAP which, not surprisingly, selects the trapezoidal numeric integration method. The trapezoidal numeric integration method is a relatively fast, accurate method, but trapezoidal
integration is not foolproof. The trapezoidal method suffers from an
annoying tendency to oscillate around the correct solution, especially
on switching circuits or long transient simulations. The Gear method
integration does not oscillate and tends to be more stable (stays close
to the exact solution) over long transient simulations, but at the
expense of longer simulation run times.

.OPTIONS MAXORD=X. The MAXORD=X parameter defines the maximum order for Gear's multiorder integration method. When a user
specifies METHOD=GEAR, SPICE uses the Gear second-order method.
Higher-order Gear methods are available, including Gear's third-,
fourth-, fifth-, and sixth-order methods. To utilize one of the higher-
order methods, the option MAXORD must be set to the highest order
which will be used. MAXORD=3 specifies the third-order Gear method,
MAXORD=4 specifies the fourth-order Gear method, MAXORD=5
specifies the fifth-order Gear method, and, finally, MAXORD=6 specifies the sixth-order Gear method.

The MAXORD=X parameter has a default value of 2 and may be set
between 2 and 6.

It should be noted, in theory, the higher-order Gear methods should
produce less error at each timepoint, thereby resulting in less total
timepoints and a faster transient simulation; due to the extra overhead
involved with the higher-order Gear methods, however, none of the
higher-order methods is significantly faster than the Gear second-
order method.[2]

For a more thorough discussion on numeric integration methods and
which type of integration to use for each circuit type, readers are
directed to Chap. 4 of this text.

.OPTIONS MU=X. The MU=X parameter was never documented in the
SPICE2G.6 User's Guide.[1] Like several other options, MU serves as
both a flag and a parameter value. The default value of MU is .5.

To understand MU, the equations for trapezoidal integration and
backward-Euler numeric integration should be compared. There is a
striking similarity between the two methods. If one carefully writes
the equations with a single, well-placed variable, the value of the
variable may be used to change the equation from trapezoidal integration to backward-Euler integration. In SPICE, MU is that variable. By setting MU to .5, the default setting, the equations reduce to
the trapezoidal integration method. By setting MU to 0, the equations
reduce to the backward-Euler integration method, and by setting

MU somewhere between 0 and .5, a blending of the two methods is used. Readers are directed to Chap. 4 for a more thorough discussion of integration methods and when to use backward-Euler or trapezoidal integration.

.OPTIONS TRTOL=X CHGTOL=X. The TRTOL=X and CHGTOL=X parameters are two of the more misunderstood parameters. Because the names end in TOL, much like RELTOL, ABSTOL, and VNTOL, many people believe these are error tolerances also. Unfortunately this is not the case. Both the TRTOL and CHGTOL options were installed during the development of the Local Truncation Error (LTE) timestep control algorithm, and both are related to the LTE algorithm only.

During the development of the LTE algorithm, the predicted value of local truncation error was found to be seven times larger than the exact value of truncation error.[2] To compensate for this discrepancy, the parameter named TRTOL was added to the equation for truncation error, and, not surprisingly, TRTOL has a default value of 7.

While TRTOL has a direct relationship on the overall stepsize used during transient analysis, increasing TRTOL increases the stepsize, while decreasing TRTOL decreases the stepsize. Because TRTOL was added to compensate for a poor estimate of truncation error, this parameter is best left at the default value of 7.

At the same time, it was observed that on certain circuits the Local Truncation Error timestep control algorithm would "lock up" by producing a step size that was infinitesimally small. This small stepsize occurred whenever the charge on a capacitor or flux through an inductor was less than the error tolerances of the program (RELTOL, ABSTOL, and VNTOL). To prevent this from occurring, a parameter named CHGTOL was added to the equations as a lower limit on capacitor charge or inductor flux. Whenever the circuit produces a maximum capacitor charge or inductor flux which is less than CHGTOL, SPICE uses the value of CHGTOL in the LTE equation to predict the next timestep. CHGTOL is simply a means of preventing the LTE timestep control method from failing.

CHGTOL has a default value of 1e-14 coulombs. Lowering this value does not significantly increase accuracy and only increases the possibility of lengthening the total transient simulation time. For these reasons, CHGTOL is best left at the default value.

Numerical pivoting options

The last two numeric control options were also the last two options added to SPICE. One of the last major changes made to the program was the addition of a modified numerical pivoting algorithm. Pivoting reorders the matrix to reduce the number of operations required to solve the matrix.

.OPTIONS PIVTOL=X PIVREL=X. The PIVTOL=X and PIVREL=X parameters both relate to the numerical pivoting algorithm in SPICE. The PIVTOL parameter defines the smallest numeric value (matrix entry) which is considered an acceptable matrix entry. If an entry is less than PIVTOL, a numerical overflow condition (such as a divide by zero) could occur. To prevent this, SPICE checks each of the matrix entries to ensure no pivot entry is less than PIVTOL. The PIVREL=X parameter defines the ratio between the largest entry in a given column of the conductance array and PIVTOL.

Numerical pivoting is a largely experimental science. Optimum pivoting is different for different circuits. The default value of PIVTOL is 1e-13, the default value for PIVREL is .001, and both were determined empirically based on 64-bit double-precision conductance values. For this reason, both PIVTOL and PIVREL should be left at their default value.

Numeric control options summary

To achieve fast, accurate, convergence simulations, SPICE users must understand and know how to set the 18 parameters which form the numeric control option group. Quality simulations require significant user input. Part of the input is supplied in the form of these 18 numeric control option parameters. Chapters 3, 4, and 5 of this text describe in detail how each parameter should be set for a given circuit. Study these chapters and become familiar with the numeric control options. Time invested in these 18 parameters will produce the greatest return in quality simulation when compared with other aspects of circuit simulation.

MOS Geometry Options

The next four parameters form the MOS geometry options. All four are applicable to MOS transistors only. The geometry of an MOS transistor gate has a large impact on the I-V and C-V characteristics of the device. Because, on any integrated circuit, many transistors share common geometries, four options were added to SPICE to reduce the effort in describing the transistor geometries in the circuit netlist.

.OPTIONS DEFL=X DEFW=X

The first two of these are the DEFL=X and DEFW=X parameters. The DEFL option describes the default gate length SPICE uses in the calculation of the transistor drain current. In many IC houses, the minimum gate length is defined by the process of fabrication, and designers modify the width of the gate to adjust the drive current of the transistor. If the DEFL=X parameter is set to the minimum allowed gate length, designers need only specify the transistor gate width on the element line. If a designer chooses to use a transistor with a gate length other than the default, simply adding the L=X description to the element line overrides the default value declared with DEFL.

The gate width parameter DEFW=X defines the default width of the transistor gate. Like the DEFL=X parameter, DEFW=X may be overridden by specifying W=X on the element line. For both DEFL=X and DEFW=X, the units of length and width are in meters. Remember to add the U (micro) scale factor if you describe these quantities in microns. Both the DEFL and the DEFW parameters default to 100 microns.

.OPTIONS DEFAD=X DEFAS=X

Just as the gate length and width may be described by default settings, so may the area of the drain and source be assigned default values. The area of drain and source are used in the capacitance calculation for the drain and source capacitance if the CJ model parameter is declared. The DEFAD=X defines the default value of the drain diffusion area and DEFAS=X defines the default value of the source diffusion area. Like the length and width parameters, both the drain area and source area are in units of meters squared.

Miscellaneous Options

The last four options fall into a category called the miscellaneous options.

.OPTIONS TNOM=X

The first of these is the TNOM=X parameter. TNOM defines the nominal temperature of analysis. In most cases this is room temperature, or the nominal junction temperature. Values for TNOM should be expressed in units of degrees Celsius. By default, TNOM is set to 27C. While this is warm for room temperature, the assumption is that, even at room temperature, a semiconductor junction is generating some heat. TNOM represents the temperature of the device junctions.

All of the temperature-dependent terms in SPICE change according to a change in temperature away from TNOM. TNOM should be the temperature used as a basis for temperature-dependent model parameter or element value extraction.

.OPTIONS CPTIME=X LIMTIM=X

The CPTIME=X and LIMTIM=X parameters are two more options which were installed when SPICE was originally developed and were added to limit the amount of computer resources a given simulation job could use.

The CPTIME=X parameter defines the maximum CPU time which could be used to simulate a circuit. Because of the continued reduction in cost of computer resources, most versions of SPICE have disabled this option by setting the default value extremely high. By default, CPTIME is set to 1 billion seconds. (For those of you who are reaching for your calculators to see how many days that translates into, let me save you some time. One billion seconds is approximately 32 years!)

Like CPTIME, LIMTIM is another safety net added to save computer resources. The LIMTIM=X parameter defines the maximum time allowed to generate the output. This parameter has been disabled in many versions of SPICE. By default, LIMTIM is set to 2 seconds.

.OPTION LVLCOD=X

The last option is named LVLCOD. The LVLCOD=X parameter was used in early CDC versions of SPICE. The default for LVLCOD is 0. If LVLCOD was set to 1, early versions of SPICE called a subroutine named CODGEN.

During the iterative procedure, with each new iteration, SPICE must solve the system equations array (the computer equivalent of performing a Gaussian elimination on a linear set of equations). SPICE solves the arrays using several nested fortran do-loops. Because fortran do-loops tend to be relatively inefficient on large, sparse matrices, the authors of SPICE decided to add a routine which would bypass the fortran do-loops and replace the solution with direct machine instructions. The routine which did this was named CODGEN. CODGEN wrote machine instruction to solve the system equations directly. While the addition of CODGEN reduced simulation time by 10–30 percent, the routine was dropped from later versions of SPICE. CODGEN was removed from SPICE for two reasons: first, because compiler efficiency and computer speed have both increased dramatically and, second, because the machine instructions were written specifically for a CDC computer. This made porting SPICE to an IBM or Cray or Vax or any other computer a nightmare. For both of these reasons, the CODGEN routine has been eliminated from most versions of SPICE.

Suggested Option Settings

For most circuits, performance may be substantially improved by resetting several of the default option values. The following are suggested settings for all circuits.

```
.OPTIONS ACCT LIST ITL1=500 ITL2=200 ITL4=40 ITL5=0 LIMPTS=10K
```

Other options which need to be set include GMIN=X, RELTOL=X, ABSTOL=X, VNTOL=X, LVLTIM=X, and METHOD='ABCD', but all of these must be set to values appropriate for your circuit. Chapters 3, 4, and 5 discuss setting specific options for different types of circuits.

Summary

The 34 options in SPICE define the information printed in the output file, the numeric algorithms used in the program, and iteration limits of the program. Learning to use and knowing how to set the proper options for a given circuit define the difference between a knowledgeable SPICE user and a novice, and the difference between fast, accurate, convergent, high-quality simulation results and garbage. SPICE is not as autonomous or reliable as many believe. SPICE simulations require significant user input to produce accurate results. The inputs SPICE requires include accurate netlist connectivity, accurate device model parameter sets and element values, and accurate option parameter settings.

References

1. *SPICE Version 2G.6 User's Guide.*
2. L. W. Nagel, *SPICE2: A Computer Program to Simulate Semiconductor Circuits,* Electronics Research Laboratory Rep. No. ERL-M520, University of California, Berkeley, 1975.

A Comparison of Vendor-Offered Simulator Options Settings

Option	SPICE2G.6 (& Rspice)	Default Settings Hspice	IS_Spice	Micro-Cap IV	Pspice
LIMPTS	201	2001	201	n/a	n/a
NUMDGT	4	4	4	n/a	4
GMIN	1pS	1pS	1pS	1pS	1pS
RELTOL	.001	.001	.001	.001	.001
VNTOL	1uV	50uV	1uV	1uV	1uV
ABSTOL	1pA	1nA	1pA	1pA	1pA
ITL1	100	200	100	100	40
ITL2	50	50	50	50	20
ITL3	4	3	4	4	n/a
ITL4	10	8	10	10	10
ITL5	5000	0	5000	1000000	n/a
ITL6	0	n/a	0	n/a	n/a
LTLTIM	2	1	2	n/a	n/a
METHOD	TRAP	TRAP	TRAP	n/a	n/a
MAXORD	2	2	2	n/a	n/a
MU	.5	.5	.5	n/a	n/a
TRTOL	7	7	7	7	n/a
CHGTOL	.01pC	.001pC	.01pC	.01pC	.01pC
PIVTOL	.01p	.001p	.001p	n/a	n/a
PIVREL	.001	.001	.001	n/a	n/a

Index

ABOUT THE AUTHOR

Ron Kielkowski developed the *Successfully Simulating Circuits with SPICE* training seminar for RCG Research. He is the primary instructor for this course and currently teaches it throughout the United States. Mr. Kielkowski also led the development of the RSPICE and RGRAPH programs for RCG Research and is the editor of the newsletter *Inside SPICE*. Previously, Mr. Kielkowski directed simulation activities at both the IC Design Center of Delco Electronics and the Advanced Development Center of the Naval Air Warfare Center.

ABOUT THE DISK

On the disk which accompanies this text are the RSPICE and RGRAPH programs. RSPICE is a PC version of Berkeley SPICE2G.6. RGRAPH is a sophisticated graphical postprocessor which may be used to view and plot simulation results. Both of these programs use extended memory and true 32-bit addressing. This makes these programs exceptionally fast and powerful.

RSPICE and RGRAPH were written by RCG Research, Inc. for use with their Successfully Simulating Circuits with SPICE training classes. Since much of the material for this text comes from these same classes, it seemed appropriate to include the programs as well. Even the simulation graphics illustrated in the text were generated with RSPICE and RGRAPH.

The RSPICE and RGRAPH programs provided with this text are free programs with a reserved copyright. Individuals may use, copy, and distribute these program free of charge. Users may wish to register their copies with RCG Research for technical support and upgrade notices as the programs evolve.

The program disk also contains problem circuits and solutions for Chaps. 2, 3, 4, 5, and 6. Readers are encouraged to simulate these circuits with RSPICE and RGRAPH as they read the text. The educational value of seeing a simulation performed both the right way and the wrong way is irreplaceable.

DISK WARRANTY

This software is protected by both United States copyright law and international copyright treaty provision. You must treat this software just like a book, except that you may copy it into a computer to be used and you may make archival copies of the software for the sole purpose of backing up your software and protecting your investment from loss.

By saying, "just like a book," McGraw-Hill means, for example, that this software may be used by any number of people and may be freely moved from one computer location to another, so long as there is no possibility of its being used at one location or on one computer while it is being used at another. Just as a book cannot be read by two different people in two different places at the same time, neither can the software be used by two different people in two different places at the same time (unless, of course, McGraw-Hill's copyright is being violated).

LIMITED WARRANTY

McGraw-Hill warrants the physical diskette(s) enclosed herein to be free of defects in materials and workmanship for a period of sixty days from the purchase date. If McGraw-Hill receives written notification within the warranty period of defects in materials or workmanship, and such notification is determined by McGraw-Hill to be correct, McGraw-Hill will replace the defective diskette(s). Send requests to:

Customer Service
TAB/McGraw-Hill
13311 Monterey Ave.
Blue Ridge Summit, PA 17294-0850

The entire and exclusive liability and remedy for breach of this Limited Warranty shall be limited to replacement of defective diskette(s) and shall not include or extend to any claim for or right to cover any other damages, including but not limited to, loss of profit, data, or use of the software, or special, incidental, or consequential damages or other similar claims, even if McGraw-Hill has been specifically advised of the possibility of such damages. In no event will McGraw-Hill's liability for any damages to you or any other person ever exceed the lower of suggested list price or actual price paid for the license to use the software, regardless of any form of the claim.

McGRAW-HILL, INC. SPECIFICALLY DISCLAIMS ALL OTHER WARRANTIES, EXPRESS OR IMPLIED, INCLUDING BUT NOT LIMITED TO, ANY IMPLIED WARRANTY OF MERCHANTABILITY OR FITNESS FOR A PARTICULAR PURPOSE. Specifically, McGraw-Hill makes no representation or warranty that the software is fit for any particular purpose and any implied warranty of merchantability is limited to the sixty-day duration of the Limited Warranty covering the physical diskette(s) only (and not the software) and is otherwise expressly and specifically disclaimed.

This limited warranty gives you specific legal rights, you may have others which may vary from state to state. Some states do not allow the exclusion of incidental or consequential damages, or the limitation on how long an implied warranty lasts, so some of the above may not apply to you.